Neo-Pre-Platonic Naturalism

Neo-Pre-Platonic Naturalism

A First-Principles Framework for Reality, Mind, and Knowledge

Eli Adam Deutscher

Neo-Pre-Platonic Press

An Imprint of the NPN System

2025

Copyright © 2025 by Eli Adam Deutscher

All rights reserved. No part of this publication may be reproduced, distributed, or transmitted in any form or by any means, including photocopying, recording, or other electronic or mechanical methods, without the prior written permission of the publisher, except in the case of brief quotations embodied in critical reviews and certain other non-commercial uses permitted by copyright law.

Neo-Pre-Platonic Press

First Edition: 2025

ISBN: 978-1-971183-00-8 (Hardcover: Standard Edition)
ISBN: 978-1-971183-03-9 (Hardcover: Library Edition)
ISBN: 978-1-971183-02-2 (Trade Paperback)
ISBN: 978-1-971183-01-5 (Ebook/PDF)

US Version: Printed in the United States of America.

Official updates and distribution: **www.neopreplatonic.com**

"The Logos holds forever, but humans always prove unable to understand it..."
Heraclitus (DK 22 B1)

"...For they pay penalty and retribution to each other for their injustice according to the assessment of Time."
Anaximander (DK 12 B1)

*To the keepers of the archives who preserved the maps,
and to the Navigators who sail the open ocean.*

Contents

List of Tables   xix

Preface: The Genesis of a System   xxi

Prologue: The Schism   xxvii

1 Introduction: The Navigator's Toolkit   1
- 1.1 The Navigator's Dilemma 1
- 1.2 The Confidence Gradient: Our Epistemic Compass. 1
- 1.3 The Navigator's Toolkit: A Lexicon of First Principles 3

2 The Anaximandrian Compass   9
- 2.1 Introduction: The First *Physiologoi* 9
- 2.2 Thales: The Principle of the Unified Source 11
- 2.3 Anaximander: The Life-Cycle of Identity 12
- 2.4 Heraclitus: The *Logos* of *Physis* 13
- 2.5 Empedocles: The Engine of Complexity 14
- 2.6 The Atomists: The Hardware of the Cosmos 15
- 2.7 Parmenides: The Architecture of the Subject 15
- 2.8 Xenophanes of Colophon: The Limits of the *Nous* 16
- 2.9 The Unified Model: *Physis* and the Emergent *Nous*. 18
- 2.10 Conclusion: The Compass Restored 20
- 2.11 The Navigator's Focus: Terms for the Socratic Turn 21

3 Socrates as a Physician   23
- 3.1 Introduction: The Misplaced Philosopher 23
- 3.2 The Pre-Platonic Project: Logos as the Standard of Health 24
- 3.3 Socrates the Physician: The Diagnosis of Cognitive Sickness . . . 25
- 3.4 The Therapeutic Goal: From Lysis to Aporia 27

3.5	Socrates as the Pre-Platonic Culmination	28
3.6	The Platonic Divergence: The Metaphysical Prosthesis.	28
3.7	Conclusion: Healing the Unexamined Soul	29

4 NPN Metaphysics   31

4.1	Introduction: First-Principles Architecture	31
4.2	First Philosophy: The Geometry of Distinction	33
4.3	The Ground: The General Zero Principle and Its Application . . .	34
4.4	The Foundation: The Primacy of the *Archē*	37
4.5	The Boundary of Knowledge: The Apeiron	40
4.6	The Logos: Lawful Structure in Nature.	42
4.7	Conclusion: The Laws of the Territory	48

5 Agency: Hormē and the Origin of Value   51

5.1	Introduction: Cosmic Laws to the Conscious Navigator	51
5.2	The Constitutive Hormē: Agency as Striving	52
5.3	Ethical Isomorphism	59
5.4	The Life-Agency Isomorphism.	60
5.5	Conclusion: The Imperative to Navigate	61

6 Architecture of the Navigator   63

6.1	Introduction: The Evolutionary Layering of the *Hormē*.	63
6.2	The Stratified Psyche: Layering of the Horme	64
6.3	Topography of Navigation: Consciousness and Reaction	70
6.4	Polarity of the *Psyche*: Cognitive Philia and Neikos	74
6.5	Conclusion: The Grounded Navigator	76

7 NPN Epistemology: The Somatic logos   79

7.1	Introduction: Somatic logos and Confidence Gradient	79
7.2	The Somatic *logos* as an Evolved Heuristic	80
7.3	The Epistemic Process: Perception to Knowledge	85
7.4	Primary Corollary: The Confidence Gradient	87
7.5	Conclusion: The Grounded Navigator	89

8 The Art of Navigation — 93
- 8.1 Introduction: Practice of Philosophical Realignment — 93
- 8.2 Induction of Dissonance: The Socratic Method — 94
- 8.3 Navigating the *Apeiron*: The Post-Collapse Phase — 96
- 8.4 *Dikē* (Δίκη): Acceptance of Causal Constraints — 97
- 8.5 Methodological Choice: *Prohairesis* in Inquiry — 99
- 8.6 *Energeia*: Active Implementation of New Model — 100
- 8.7 Conclusion: The Integrated Model — 103

9 The Platonic Synthesis — 107
- 9.1 Introduction: From *Kathairesis* to Synthesis — 107
- 9.2 Confronting the Void — 108
- 9.3 The First Misstep: *Anamnesis* — 109
- 9.4 Metaphysical Flight: *Phaedo* and Fear of *Apeiron* — 110
- 9.5 Motivational Engine: *Symposium* and *Eros* of Escape — 112
- 9.6 The Political Synthesis: The *Republic* — 113
- 9.7 Conclusion: The *Republic* as a Monumental Warning — 116

10 The Navigator and the Labyrinth — 117
- 10.1 Introduction: The Two Platos — 117
- 10.2 The Great Unbuilding (*Lysis*) — 119
- 10.3 The New Foundation (*Synthesis*) — 123
- 10.4 The Pragmatic Reckoning — 127
- 10.5 Conclusion: Unfinished Symphony and Conductor — 130

11 The Aristotelian Impasse — 135
- 11.1 Introduction: The Completion of the Immanent Turn — 135
- 11.2 The Synchronic Summit: Mapping the Static *Archē* — 136
- 11.3 From Static Substance to Diachronic Becoming — 141
- 11.4 Political Consequence: Cost of Flattened Reality — 143
- 11.5 NPN Ideal: Society as Emergent Property — 145
- 11.6 Conclusion: The Harbor and the Ocean — 147

12 NPN Politics — 149
- 12.1 Introduction: The *Polis* as a Natural Phenomenon — 149

12.2	The NPN Analysis: The *Physis* of the *Polis*	150
12.3	Two Archetypal Forms: Poles of Social Tension	152
12.4	The NPN Prescription: The Second-Best State	155
12.5	The Ideal: The Navigator Polis	158
12.6	Conclusion: The North Star	162

13 NPN Aesthetics — 165

13.1	Introduction: Art as a Natural Human Behavior	165
13.2	The Functional Purpose of Art: *Mimesis*	166
13.3	The NPN Theory of Beauty: Aesthetic Judgment	171
13.4	Conclusion: The Final Affirmation	173

A The General Zero Principle — 175

A.1	The Problem of Infinite Regress	175
A.2	Circular Reasoning as Infinite Regress	176
A.3	The GZP Solution: The Indeterminate Foundation	176
A.4	Historical Recognition of the Problem	178
A.5	Modern Formalization: From Spinoza to Spencer-Brown	180
A.6	The *Apeiron* and the Meta-Regress	182
A.7	Application to NPN First Principles	182
A.8	The Ultimate Justification	184
A.9	Conclusion: The First First-Principle	184

B First Principles of NPN — 187

B.1	Introduction: The Core Principles of NPN	187
B.2	The Ground: The Precondition of Existence	188
B.3	The Foundation: The Nature of Reality	189
B.4	The Boundary: The Limits of Knowledge	191
B.5	The Bridge: The Principles of Relation	192
B.6	The Navigator: The Emergent Agent	193
B.7	Theorems: Derived Doctrines	195

C Anaximander and the Zero Principle — 203

C.1	The First Philosopher of Relation	203
C.2	Why Aristotle Could Not See It	207

CONTENTS

C.3	Conclusion: Recovering the First Metaphysics.	208

D Plato and the Zero Principle — 211

D.1	Introduction: Plato's Unrecognized Foundation	211
D.2	The Middle Dialogues: Building the Determinate Background. . .	211
D.3	The Late Dialogues: The System Under Stress	213
D.4	Plato's Struggle Towards General Zero Principle.	215
D.5	Why Plato Matters for NPN	215
D.6	Conclusion: Plato's Tragic Genius	216

E Aristotle and the Zero Principle — 217

E.1	Introduction: The Systematic Correction	217
E.2	Formalizing the Ground: Potentiality and Actuality	218
E.3	The Biological Turn: Mapping the Results of Process.	219
E.4	The GZP Critique of the Unmoved Mover.	220
E.5	The Four Causes and the Relational Web	221
E.6	Aristotle's Legacy: The Data and the Framework	221
E.7	Conclusion: Aristotle as Bridge and Limit.	222

F The Confidence Gradient — 225

F.1	Introduction: The Allure and Peril of the Binary.	225
F.2	The Functional Anatomy of the Confidence Gradient	226
F.3	Destruction of Certainty is Acquisition of Power	227
F.4	From Nihilism to Navigational Vigor	228
F.5	The Popperian Protocol: Falsification as Navigation	229
F.6	A Scientific Exemplar: The Heisenberg Uncertainty Principle . . .	232
F.7	Derivation: Necessity of Confidence Gradient	232
F.8	Conclusion: Certainty is a Cage, The Gradient is the Compass . .	235

G The Navigator Protocol — 237

G.1	Introduction: The Logic of Practice.	237
G.2	Phase I: The Diagnosis (Breaking the Shell)	237
G.3	Phase II: The Reconstruction (Aligning with Reality)	239
G.4	Phase III: The Execution (Action and Mastery).	240
G.5	The Cybernetic Loop: The Popper Protocol	241

G.6	Summary Table: The 11 Stages of the Navigator	243
G.7	Conclusion: The Architecture of Alignment	244

H The Universal Grammar: Daoism and NPN — 247

H.1	Introduction: Vocabulary for Reality's Rhythm	247
H.2	Divergence: Structural Analysis vs. Navigation	253
H.3	The Political Corollary: Architecture vs. Cultivation	255
H.4	The NPN Synthesis: Merging East and West	255
H.5	Conclusion: The Universal Grammar of Identity	256

I The Somatic Dhamma: Buddhism and NPN — 257

I.1	Introduction: An Unexpected Convergence	257
I.2	Diagnosis: *Dukkha* as Rebellion Against One's Own Lineage	258
I.3	The Prescription: The Eightfold Path	259
I.4	The Mechanism: Surrender to the *Somatic logos*	262
I.5	Conclusion: Intuitive Alignment vs. First-Principles Logic	263

J Objections & Rebuttals — 265

J.1	Introduction: Strengthening the System	265
J.2	The Foundation: Objections to Reality and Morality	265
J.3	The Bridge: Objections to Mind and Knowledge	270
J.4	The Boundary: Metaphysical and Theological Objections	273
J.5	Conclusion: The Resilience of Diachronic Navigator	275

K First Philosophy: The Boundary Condition — 277

Introduction: The Empty Starting Point	277
Part I: The Geometry of Distinction	278
Part II: From Geometry to Cosmos	282
Part III: First Philosophy as Navigation	286
Conclusion: The Boundary Condition	291

Ode to the Giants — 293

Glossary — 297

Notes on the Method — 309

Acknowledgments	**311**
References	**313**
Index	**327**
About the Author	**337**
Production Note	**339**

List of Tables

1.1	The Navigator's Toolkit: Core Lexicon	3
1.2	The Fracture of Meaning Across Scales	5
2.1	The Toolbox of the Physician: Key Terms for Chapter 3	22
4.1	The Core Drivers: *Philia* and *Neikos*	46
5.1	The Structure of the *Psyche*: Lexicon for Chapter 6 . . .	62
6.1	The Stratified Psyche: Evolutionary Layers of Agency .	68
6.2	Topography of Navigation: Mapping Consciousness . .	74
6.3	The Internal Polarity of the *Nous*	76
6.4	The Epistemic Chain: NPN Terms for Chapter 7	77
7.1	The Epistemic Chain: From Data to Model	87
7.2	The Navigator's Path: NPN Terms for Chapter 8	90
8.1	The Methodological Paradox: Tool vs. Goal	100
8.2	The Navigator's Path: from *Doxa* to *Eudaimonia*	105
10.1	Plato's Collection and Division: The NPN Synthesis . .	125
10.2	The Two Platos: The Shift from Architect to Navigator	132
11.1	The Synchronic Flattening: Mapping Causes to Forces .	140
11.2	The Great Philosophical Divergence	148
12.1	The Political Archetypes: Order vs. Adaptation	155
13.1	The Artist's Prohairesis: Two Modes of Creation	169
B.1	Summary of First Principles, Corollaries, and Theorems	201
C.1	Anaximander's Four-Step Cycle in NPN Terms	206
C.2	Anaximander's Relational Ontology	207
G.1	The Navigator Protocol Mapped to First Principles . . .	243
H.1	The Methodological Divergence: Structure vs. Flow . .	254
I.1	The Noble Eightfold Path as an NPN Protocol	261

Preface: The Genesis of a System

"I realized that it was necessary, once in the course of my life, to demolish everything completely and start again right from the foundations if I wanted to establish anything at all in the sciences that was stable and likely to last." —René Descartes[1]

A Decade of First-Principles Inquiry

What follows is not merely the record of a philosophical recovery mission, but the culmination of a personal and intellectual odyssey. This section is an account of the method behind this work—a method that was not chosen, but discovered through a decade-long engagement with reality itself.

The journey began over ten years ago, not in a library, but in the laboratory of my own mind. I initiated a deliberate, unflinching project: to transform my own thought process into a rigorously first-principles-based instrument, one that would admit zero *Doxa*—no unexamined belief, no inherited assumption.[2] I began to investigate reality itself, testing the limits of what a conscious mind could learn about its own constitution and the world it inhabits.

---

1. René Descartes, *Meditations on First Philosophy*, "First Meditation," in *The Philosophical Writings of Descartes*, vol. 2, trans. John Cottingham et al. (Cambridge: Cambridge University Press, 1984), 12.

2. This axiomatic approach is methodologically indebted to Aristotle's conception of first principles as the necessary foundation of demonstrative science. See Aristotle, *Metaphysics* Γ.3, 1005b5–34.

The Ethical Foundation:
First Principles from the Bottom Up

Remarkably, the system did not emerge in the traditional philosophical sequence. I arrived at the core ethical framework—the objective derivation of value from the constitutive *Hormē*—nearly five years before fully systematizing the epistemology and metaphysics. This was no top-down deduction, but a bottom-up discovery: observing that our lower psychological faculties (*Orexis* and *Thymos*) represented the accumulated results of *Philia* (cohesion) and *Neikos* (separation) operating across deep time.

The insight was both simple and revolutionary: what we call "value" is not an arbitrary assignment, but the functional relationship between a system's states and the successful expression of its constitutive striving within a lawful reality. This insight had become so fundamental to my operational worldview that it required a deliberate effort to step back and articulate it formally as a First Principle.

The Conceptual Awakening:
From Vague Intuitions to Precise Lexicon

During this fertile period, the core architecture of NPN existed in my mind with structural integrity but linguistic poverty. I operated with clear concepts that were impossible to communicate coherently. The *Apeiron* was "the infinite," the *Archē* was "Objective Reality," and the *Logos* was the "framework of reality." *Philia* and *Neikos* were described through clumsy circumlocutions about "natural selection," "aiding in survival," or "being selected against."[3]

The breakthrough came when I realized that the Presocratics had already forged the precise vocabulary I needed. Their lexicon was designed for this very purpose—to describe first principles without the accumulated baggage of millennia of philosophical reinterpretation. *Logos* perfectly captured the operational structure of reality I had discerned. *Apeiron*,

---

3. The diachronic and selective logic underpinning this system is inspired by Darwin's principle of natural selection as a creative, iterative process. See Charles Darwin, *On the Origin of Species* (London: John Murray, 1859), chap. 4.

while not perfect, was the closest analogue to the 'infinite/permanently unknowable' I had conceived. Engaging with them in depth was a revelation: I found that the causes and processes in my head were the very phenomena they had strained to articulate.

The Necessary Emergence: From Value to Navigator

As the ethical foundation solidified, a crucial deduction emerged: from objective value as our inherent guiding principle evolved over deep time, a *Logistikon* MUST emerge for anything more complex than a pure *Orexis* organism to exist. There must be a conductor to orchestrate the lower functions in navigating social interactions (*Thymos*) and reality itself. Without one, there is nothing to process the feedback signals from the *Thymos* into action—you need a processor to process data. This is a non-negotiable requirement for anything higher than reactive action (*Orexis*).

As *Thymos* emerged, the *Logistikon* must emerge with it to process the feedback of the *Thymos* into coherent social interactions. This realization —that the *Nous* is a necessary navigator—brought with it the crucial corollary: the map (our modeling of reality) could never be certain that it was the territory. *Epistēmē* was necessarily a gradient, not a binary, and the feeling of binary choice is the cognitive collapsing of our four-dimensional world into a three-dimensional snapshot to facilitate decision-making.

The Socratic Method: From *Aporia* to *Synthesis*

The final piece fell into place when I recognized that the process for analyzing reality itself was already embedded in the Socratic method. Socrates provides the initial steps for reaching *Aporia*—the essential clearing of cognitive ground—while NPN provides the *Synthesis*, the systematic framework for rebuilding on first principles. The *Elenchus* is not mere skepticism, but the necessary induction of productive perplexity that precedes genuine understanding.

I am now convinced this iterative, self-correcting process is the only possible path to a coherent, first-principles understanding. The *Navigator* had to navigate before the entire territory of existence—the *Archē* and the *Apeiron*—could be constituted into a coherent whole. This was not an academic exercise; it was a slow, methodical grappling with all branches of philosophy, often in parallel, treating the entire history of thought as a single, vast system that must be rendered non-contradictory.

The System Emerges: From *Orexis* to *Archē*

This process of deduction—from the ethical imperative rooted in the *Hormē*, to the necessary emergence of the *Logistikon*, to the reclamation of the Pre-Socratic lexicon, and finally to the formalization of the Socratic crisis as a navigational protocol—revealed itself to be a single, unified structure. What began as a personal project of cognitive hygiene evolved, through relentless iteration and reality-testing, into a complete philosophical meta-structure: **Neo-Pre-Platonic Naturalism**.

Then, as I awaited the proof copies, came the final realization: the **Zero Principle**—that identity requires contrast. With this principle, the framework was complete. And through its lens, a historical insight came into sharp focus: Anaximander's *Apeiron* (the Boundless) and *Dikē* (Cosmic Justice) were not primitive materialism, but a sophisticated relational ontology—one that aligns precisely with the Zero Principle formalized here.[4] Sometimes, to see the past clearly, you must first rebuild the lens through which you look.

This system did not descend from the sky of abstract speculation. It ascended from the ground of lived inquiry—from the observation that a mind committed to first principles cannot tolerate contradiction, not only within its own beliefs, but within the entire inherited edifice of Western thought. The "history of philosophy" ceased to be a chronology of disparate schools and became a diagnostic map of humanity's collective navigation, marked by moments of profound insight and catastrophic wrong turns. The Pre-Socratics had charted the coastline; Socrates taught us how to sail; Plato built a glorious but unseaworthy ship; Aristotle perfected the

---

4. For the complete etymological and philosophical analysis of Anaximander's fragment in relation to the Zero Principle, **see Appendix C: Anaximander and the Zero Principle: The Life-Cycle of Identity**.

harbor. And we have been living in that harbor ever since, admiring its beauty but forgetting the open ocean.

This book is the log of that ocean voyage. It is the synthesis of a decade's navigation, offered not as dogma, but as a tested map and a restored compass. Its First Principles are not articles of faith, but the non-negotiable coordinates that emerge when one commits fully to the **Diachronic Turn** —to seeing reality, mind, and knowledge as processes in time, as *Becoming* rather than static *Being*.

The system did not descend from the sky of abstract speculation. It ascended from the ground of lived inquiry, from the observation that a mind committed to first principles cannot tolerate contradiction—not only within its own beliefs, but within the entire inherited edifice of Western thought. The "history of philosophy" ceased to be a chronology of disparate schools and became a diagnostic map of humanity's collective navigation, marked by moments of profound insight and catastrophic wrong turns. The Pre-Socratics had charted the coastline; Socrates had taught us how to sail; Plato built a glorious but unseaworthy ship; Aristotle perfected the harbor. And we have been living in that harbor ever since, admiring its beauty but forgetting the open ocean.

This book is the log of that ocean voyage. It is the synthesis of a decade's navigation, offered not as dogma, but as a tested map and a restored compass. Its First Principles are not articles of faith, but the non-negotiable coordinates that emerge when one commits fully to the **Diachronic Turn** —to seeing reality, mind, and knowledge as processes in time, as *Becoming* rather than static *Being*.

A Note to the Reader: You Are the Navigator

You now hold the completed system. But a map is not a journey, and a compass does not steer the ship. The value of NPN lies not in passive agreement, but in active application. Its ultimate test is not in its internal consistency—though that has been rigorously pursued—but in its utility. Does it help you navigate? Does it resolve the old schisms in your own thinking? Does it provide a clearer bearing in the moral and epistemic fog of our age?

I invite you, therefore, to engage not as a spectator, but as a collaborator. Subject these principles to your own *Elenchus*. Test them against the *Logos* as you encounter it. Use them to diagnose the *Doxa* in your own life and in the world around you. This system was forged in the solitary workshop of one mind, but it is meant for the common toolkit of all who choose to think clearly and live deliberately.

We stand at a peculiar moment in history, equipped with tools of unprecedented power and confronted with crises of unprecedented scale. We have built prosthetic *Noes* (plural *Nous*) that can access all human thought, yet we lack a grounded ethics to guide their use. We have more information than ever, yet less shared understanding. The old maps are tearing at the seams.

It is time for new navigation.

This book is offered as a contribution to that project—not as the final word, but as a foundational one. The *Archē* is. The *Logos* is structured. The *Nous* is emergent. The *Hormē* strives. From these simple, inescapable starting points, an entire world of meaning and method unfolds. The rest is up to you.

—Eli Deutscher
2025

Prologue: The Schism

"Thoughts without content are empty, intuitions without concepts are blind."
—Immanuel Kant, *Critique of Pure Reason*, A51/B75

"Thinking substance and extended substance are one and the same substance, comprehended now under this attribute, now under that."
—Baruch Spinoza, *Ethics*, Part II, Proposition 7, Scholium

The Conceptual Crisis: The Schism at the Heart of the West

Modern philosophy and public discourse are characterized by a deep fragmentation. This fragmentation is not new; it is the symptom of a 2,500-year-old philosophical civil war between two seemingly irreconcilable camps: Idealism and Materialism.

- **The Idealist Intuition:** That mind, consciousness, and abstract logic are the primary reality—too profound and necessary to be mere byproducts of matter.

- **The Materialist Intuition:** That the physical world is the only reality—and that consciousness must therefore be an illusion or a trivial accident.

This schism is the root of our deepest dualisms: Mind vs. Body, the One vs. the Many, Being vs. Becoming. It is a philosophical wound that has never healed, forcing every thinker to take a side in a debate where both sides seem partially right and catastrophically wrong. We have been trying to understand a dynamic, unified reality through a static, fractured lens.[5]

The Thesis: A Diachronic Turn to First Principles

This work proposes that the path out of this crisis lies in a fundamental shift in perspective: the **Diachronic Turn**. The Idealism/Materialism debate is a quintessential synchronic problem, analyzing reality in a timeless snapshot. Neo-Pre-Platonic Naturalism (NPN) initiates a diachronic turn, arguing that reality (*Archē*), mind (*Nous*), and knowledge (*Epistēmē*) are dynamic, evolutionary processes that can only be understood across time.[6]

This diachronic framework is the master key that unlocks the old prison:

- It allows us to see the *Logos* not as a static, transcendent law (Idealism) or a mere descriptive summary (Materialism), but as the **operational grammar of a cosmic process**.

- It allows us to see the human mind not as a spectral outsider (Idealism) or a determined epiphenomenon (Materialism), but as an **evolved navigator** whose internal logic is a hard-won reflection of the external world.

- It resolves the central schism by showing how a primary, physical *Archē* can, through diachronic processes, give rise to a conscious *Nous* whose complex products were mistakenly seen as evidence for a separate, ideal realm.

---

5. For a foundational 20th-century diagnosis of Western philosophy as a civil war between objectivism and subjectivism, see Edmund Husserl, *The Crisis of European Sciences and Transcendental Phenomenology*, trans. David Carr (Evanston, IL: Northwestern University Press, 1970), 3–18.

6. This critique echoes Heidegger's diagnosis of the "forgetting of Being" in favor of static presence. See Martin Heidegger, *Being and Time*, trans. John Macquarrie and Edward Robinson (New York: Harper & Row, 1962), 21–25.

The result is a unified, naturalistic system that does not explain away the human experience, but grounds it in the lawful, generative processes of the cosmos. This book invites the reader to stop inspecting static snapshots and begin navigating the dynamic reality we actually inhabit.

Chapter 1

Introduction: The Navigator's Toolkit

"Give me a place to stand, and I will move the earth."
— Archimedes[1]

1.1 The Navigator's Dilemma

We stand at the shore. The Prologue has diagnosed the ancient schism and pointed to the open ocean—the Diachronic Turn. The Preface has recounted the personal voyage that mapped it. But a destination is not a journey, and a diagnosis is not a cure.

To sail, we need two things: **a compass** to tell direction amid uncertainty, and **a clean chart** whose symbols we can trust. Without the first, we drift; without the second, we mistake our own drawings for the coastline.

This chapter provides both: the **Confidence Gradient** as our epistemic compass, and the **Lexicon of First Principles** as our cartographic key. Together, they form the essential toolkit for any mind attempting the Diachronic Turn—the practical means by which we stop inspecting snapshots and begin navigating the current.

1.2 The Confidence Gradient: Our Epistemic Compass

NPN relies on *graded* epistemic justification rather than claims of absolute certainty. This is not a stylistic preference, but a structural consequence of the human condition. We are finite, time-bound, and embodied

---

[1]. Quoted by Pappus of Alexandria, *Synagoge*, Book VIII, 1060.

agents whose understanding of the world unfolds through diachronic interaction.[2] No human being has access to a complete, external, or timeless vantage point on reality. Under such conditions, **epistemic certainty is impossible in principle.**[3]

For this reason, NPN treats "knowledge" not as infallible belief but as **confidence that strengthens or weakens across time** in response to experience, evidence, and the structured constraints of the world. This dynamic process—what NPN calls the **Confidence Gradient**—provides the only rational standard for action and belief available to finite agents. It does not aim to eliminate uncertainty; it provides a principled way to navigate it.

The Confidence Gradient establishes **clear limits** on what can be considered justified, actionable knowledge. Throughout this work, every claim remains strictly within these limits: no appeals to certainty, no unrestricted metaphysical speculation, and no requirements that exceed what humans can actually know. Readers seeking the complete formal derivation of the Confidence Gradient—including its necessity, structure, and implications—will find it in **Appendix F**. This section states the epistemic boundaries within which NPN operates.

Within those boundaries, NPN makes the most honest and coherent use of what is available to us: finite cognition, diachronic experience, and the structured world in which we act. The chapters that follow proceed entirely within these constraints.

---

2. Xenophanes' fragments represent a crucial step in this forging, challenging anthropomorphic theology and advancing a proto-naturalistic epistemology based on gradual human inquiry. See James H. Lesher, *Xenophanes of Colophon: Fragments* (Toronto: University of Toronto Press, 1992), 150–169, esp. fragments B18 and B34.

3. For a pragmatic rejection of the quest for absolute, spectator-model certainty, see John Dewey, *The Quest for Certainty: A Study of the Relation of Knowledge and Action* (New York: Minton, Balch & Company, 1929), chap. 4. Dewey argues that knowledge is validated through experimental action within a contingent world, not through correspondence to fixed antecedents.

1.3 The Navigator's Toolkit: A Lexicon of First Principles

A system of first principles requires a language of first principles. Modern philosophy often operates with a lexicon that is a palimpsest of translations, interpretations, and cultural shifts. To use it is to immediately enter a contested and muddied semantic field, doomed to fight definitional battles before the substantive argument can even begin.

This work therefore builds with a reclaimed Greek vocabulary. This is not an academic affectation, but a practical necessity. The Pre-Platonic philosophers forged the first concepts for the fundamental principles of nature. The words they crafted were designed to bear this specific, immense conceptual weight.

We use them as precise technical terms—the clean, foundational components of this meta-structure. The reader is asked to treat these terms as they would specialized variables in a scientific formula. A full glossary is provided in the appendix, but the core toolkit is introduced here:

Table 1.1   The Navigator's Toolkit: Core Lexicon

NPN Term	Role	Technical Definition
Archē (Ἀρχή)	Foundation	The sole, fundamental, objective, causal system of physical reality.
Physis (Φύσις)	Territory	The essential, emergent nature and origin of all things; the *Archē* understood as a dynamic, emergent process (*Hylē*-as-structured-by-*Logos*).
Apeiron (Ἄπειρον)	Boundary	The fundamentally and permanently unknowable context for the *Archē*. The limit of the *Logos*.
Logos (Λόγος)	Structure	The inherent, discoverable, constitutive law and order of the *Archē*.

NPN Term	Role	Technical Definition
Hylē (Ὕλη)	Substrate	The primordial, unformed, and dynamic potentiality that underlies all phenomena; the receptive principle of matter.
Philia (Φιλία)	Union	The cosmic force of attraction, cohesion, and synthesis.
Neikos (Νεῖκος)	Separation	The cosmic force of repulsion, division, and differentiation.
Nous (Νοῦς)	Navigator	The emergent faculty of consciousness; the evolved capacity for model-building and understanding.
Hormē (Ὁρμή)	Engine	The constitutive, non-negotiable impulse to strive and persist in one's own being (Agency).

But why these specific terms? Why Greek? The answer lies not in tradition, but in ontological necessity.

First, the Process in "Nuanced" English:

A human—animated by a constitutive striving to persist—developed a mind. The minds that were most effective at mapping objective reality using their embodied, evolution-sculpted logic were able to harness the cosmic forces of attraction and cohesion to endure the deep-time, winnowing pressures of division and selection. Their embodied logic mirrored—because it was shaped by—the lawful structure of reality itself, this is why our embodied logic feels *a priori*. Our constitutive striving bound us together through cohesive social bonds, giving us our deep-time, adaptive social intelligence—the strategic calculator of loyalty, status, and coalition—that enabled our dominance.

Even at its best, this description is cumbersome, vague, and conceptually fractured.

Now, the Same Process in Its Native Language:

A human, an organism with *Hormē*, developed a *Nous*. The *Noes* (plural *Nous*) that were best at mapping the *Archē* using their *Somatic logos* were able to harness the *Philia* of the *Logos* to persist through the deep-time forces of *Neikos*. Their *Somatic logos* mirrored the *Logos* of the *Archē*—and that is why our logic feels *a priori*. Our *Hormē* brought us together through *Philia* and gave us our deep-time evolved *Thymos*, that allowed us to become the dominant species.

The difference is not stylistic—it is architectural.

What English Loses—And Why It Fails Miserably:

When we attempt to maintain both nuance and consistency across scales, English collapses. Consider the cosmic-to-social continuity of forces:

Table 1.2   The Fracture of Meaning Across Scales

Scale	English Attempt	What English Loses	Greek Term
Cosmic	Fundamental force of attraction/cohesion	Treated as a metaphor, not a first principle	*Philia* (Φιλία)
Evolutionary	Inheritance, symbiosis, cohesion	Splits into separate biological concepts	*Philia* (Φιλία)
Cognitive	Drive to cohere, model-building	Psychologized, divorced from cosmic force	*Philia* (Φιλία)
Social	Cooperation, trust, alliance	Sociologized, severed from root	*Philia* (Φιλία)
Cosmic	Fundamental force of repulsion/division	Reduced to "entropy" or "competition"	*Neikos* (Νεῖκος)

Scale	English Attempt	What English Loses	Greek Term
Evolutionary	Selection, variation, competition	Separated from cosmic and cognitive levels	*Neikos* (Νεῖκος)
Cognitive	Doubt, differentiation, critical analysis	Treated as mere skepticism	*Neikos* (Νεῖκος)
Social	Conflict, competition, individuation	Seen as purely political	*Neikos* (Νεῖκος)

The same fracture occurs with other terms. **English lexicalizes by domain, not by first principle.** To describe the same force operating at cosmic, evolutionary, cognitive, and social levels, you must either:

1. Use different words and lose the continuity, or

2. Strain a single English word until it becomes a vague metaphor.

The result is not just imprecision—it is **conceptual disintegration**. You cannot articulate a **fractal ontology** with a language that insists on departmentalizing reality.

Why This Lexicon Is Non-Negotiable:

The Diachronic Turn requires seeing reality as a **unified, scale-invariant process**. The Greek lexicon does not "translate" modern ideas; it **preserves the constitutive architecture of reality** that English has fragmented.[4] Each term is a **fractal key**—the same conceptual shape that fits every lock, from particle to person to polity. To think diachronically is

---

4. This aligns with Martin Heidegger's famous contention that Greek is the only language truly capable of philosophical thought because its very grammar preserves the original encounter with Being (*Sein*). See Martin Heidegger, *Introduction to Metaphysics*, trans. Gregory Fried and Richard Polt (New Haven: Yale University Press, 2000), 60–64. See also Charles H. Kahn, *The Verb 'Be' in Ancient Greek* (Indianapolis: Hackett Publishing, 2003), who demonstrates how the Greek verb *einai* seamlessly unites existence, truth, and predication in a way modern languages do not.

CHAPTER 1. INTRODUCTION: THE NAVIGATOR'S TOOLKIT

to think with first principles, and to think with first principles, we need a language that was forged for that very purpose.[5]

The following chapter is the record of that philosophical recovery mission, documenting how thinking with the first architects revealed that their fragments are not primitive precursors, but the recovered blueprints for a coherent and powerful metaphysical framework.

---

[5]. A Note on the Prologue's Epigraphs: The reader may now recognize why the epigraphs from Kant and Spinoza were left standing without immediate commentary. They represent, in distilled form, the fractured lexicon of modern philosophy: Kant's "empty thoughts" and "blind intuitions" describe a mind severed from a knowable *Archē*, while Spinoza's "one substance" gestures toward a unified *Physis* but lacks the dynamic, diachronic vocabulary of Philia and Neikos. The language of NPN does not reject these insights; it provides the constitutive connective tissue—the very terms the Pre-Platonics forged before the schism began—that allows these fractured truths to be reconciled into a stable, scale-invariant whole.

Chapter 2

The Anaximandrian Compass and the Origins of Thought

"The Non-Limited is the original material of existing things; further, the source from which existing things derive their existence is also that to which they return at their destruction, according to necessity." — Anaximander[1]

"It is the greatest thing, to know the Logos which holds all things together through all things." — Heraclitus[2]

2.1 Introduction: The First *Physiologoi*

The philosophical edifice of the West is built upon a foundation it has largely forgotten. We trace our lineage back to Socrates and Plato, to the triumphant turn toward the human *Psyche* and the transcendent Forms. But in doing so, we have relegated their predecessors to the role of curious, half-mythical precursors—brilliant but naive, stumbling in the dawn light. This is not just a historical error; it is a categorical one. The Presocratics were not preparing the ground. **They were the first *physiologoi***

---

1. Anaximander, DK 12 B1. This fragment establishes the *Apeiron* not merely as "infinite stuff," but as the *metaphysical ground* of all Becoming—the conservation law that governs the life-cycle of identity. In G. S. Kirk, J. E. Raven, and M. Schofield, *The Presocratic Philosophers: A Critical History with a Selection of Texts*, 2nd ed. (Cambridge: Cambridge University Press, 1983), 105–8.

2. The fundamental axiom of NPN epistemology: authority resides in the objective structure (*Logos*) of the *Archē*, not in the subjective rhetoric of the speaker. See Heraclitus on the *Logos* (DK 22 B50) in G. S. Kirk, J. E. Raven, and M. Schofield, *The Presocratic Philosophers: A Critical History with a Selection of Texts*, 2nd ed. (Cambridge: Cambridge University Press, 1983), 194–96.

(φυσιολόγοι)—"those who give an account of *Physis*"—the pioneering cartographers of dynamic, emergent reality itself.³

This chapter documents a philosophical recovery mission. It is the record of a live dialogue, a thinking *with* these first architects, which revealed that their fragments are not museum pieces but the foundational pillars for a coherent and powerful metaphysical framework. The path out of our modern philosophical crises—the skepticism of Hume, the specter of determinism, the abyss of moral relativism—does not lie ahead, but behind, in the primordial insights of Thales, Anaximander, Heraclitus, Xenophanes, Empedocles, and Parmenides.⁴

They provided the pieces. What follows is the synthesis.

2.1.1 The Paradigm Shift: From History to Collaboration

The standard history of the Presocratics is a story of progress: from Thales' simple "water" to the complex systems of the Pluralists, a linear march toward the sophistication of Plato. They are presented as thinkers who *posed* problems, not ones who *solved* them.

The shift that occurred in this dialogue was a move from learning *about* them to thinking *through* them.⁵ This was not an exercise in exegesis, trying to decipher what they "really meant." It was an act of philosophical collaboration across 25 centuries. The goal was not to memorize their answers, but to understand the *Physis* they were pointing to, often without

---

3. Contra the Aristotelian dismissal of the Milesians as primitive materialists. For a contemporary re-assessment of Anaximander that aligns with the NPN view of him as a foundational thinker of dynamic, self-regulating cosmic processes, see Andrew Gregory, *Anaximander: A Re-assessment* (London: Bloomsbury Academic, 2016), 43–83.

4. This chapter prioritizes philosophical coherence over strict Apollodoran chronology. NPN treats the Presocratic corpus as a "holographic" system of thought rather than a linear historical progression.

5. This hermeneutic of collaborative reconstruction finds a parallel in Gábor Betegh's interpretation of the Derveni Papyrus, which shows how early Greek cosmological and theological speculation was a live, integrative intellectual project. See Gábor Betegh, *The Derveni Papyrus: Cosmology, Theology and Interpretation* (Cambridge: Cambridge University Press, 2004), 278–310.

the precise language to describe it. They were not giving us doctrines to believe, but fundamental axes of reality to contemplate.

The core realization was this: they were creating the conceptual vocabulary for existence itself. Our task was to listen not to their specific words, but to the profound hum of the truths they were straining to articulate.

What follows is not a chronological history, but a systematic reconstruction. Each thinker identified a fundamental piece of the cosmic puzzle. Our synthesis connects these pieces into a coherent whole.

2.2 Thales: The Principle of the Unified Source

The Insight: Thales of Miletus looked past the chaotic pantheon of gods and proposed a radical idea: there is a single, natural principle—*Archē* (Ἀρχή)—underlying all things.[6] His candidate, water, is almost irrelevant. His world-shaking contribution was the act itself—the assertion that the cosmos is a unified, intelligible system accessible to human reason.[7] He was straining to identify the fundamental *Hylē* (Ὕλη)—the dynamic, undifferentiated physical potentiality that is the raw material of the cosmos.[8]

The Synthesis: Thales established the first rule of the philosophical game: seek a unified foundation. He moved the conversation from *mythos* to *logos*. While his specific answer was a placeholder, the question—"What is the fundamental nature of things?"—is the engine of all scientific and philosophical inquiry. He is the founder not of a theory, but of a method.

---

6. The term *Archē* was later coined by his student Anaximander, but the conceptual move—positing a single material source—belongs to Thales.

7. For Thales' foundational role in the shift from *mythos* to *logos*, and the *secularization of causality* (removing the gods from the physical chain of events), see Kirk, Raven, and Schofield, *The Presocratic Philosophers*, 76–99.

8. The term Hylē (Ὕλη) is used here proleptically. While Thales was attempting to isolate the material *Archē*, the precise technical definition and systematization of *Hylē* as the material cause was a later innovation by Aristotle (see section 11.2.1). The Pre-Platonics, lacking this precise lexicon, were pointing toward a concept their successors would later articulate.

2.3 Anaximander: The Life-Cycle of Identity

The Insight: Anaximander, Thales' student, saw that no determinate element could serve as the ultimate source of all opposites. He instead posited the *Apeiron* (Ἄπειρον)—**"not-bounded"**—the indeterminate ground from which all determinate things emerge and to which they return.[9] This was not a substance, but a **relational precondition**: the necessary contrast-field against which bounded identity becomes possible.

His surviving fragment reveals a complete metaphysics in four steps:

1. **The Boundless Ground (*Apeiron*)** — the indeterminate complement required for any determinate system.
2. **The Cut of Individuation (*Adikia*)** — the "injustice" of being a bounded form, cut out from the boundless.
3. **The Measure of Distinction (Time)** — the duration a form can maintain its contrast against dissolution.
4. **The Re-Cut of Dissolution (*Dikē*)** — the force that severs the form and re-integrates it into the *Apeiron*.

The Synthesis: Anaximander was not merely describing a "generative principle." He was **mapping the life-cycle of identity itself**. His system is the first articulation of what NPN formalizes as the **Zero Principle**: identity requires contrast; contrast is temporary; time measures its persistence.[10]

The *Apeiron* is the **indeterminate complement** without which no determinate thing could be. *Dikē* is not "cosmic justice" in a moral sense, but the **ontological enforcement of contrast-temporality**—the cosmic scissor that both allows and undoes individuation.

---

9. See DK 12 B 1, preserved in Simplicius, *Commentary on Aristotle's Physics*, 24.13. NPN utilizes the *Apeiron* strictly as the 'Boundary Condition' of the system—the indeterminate ground—rejecting later Neoplatonic interpretations of it as a mystical One or Deity. See also Charles H. Kahn, *Anaximander and the Origins of Greek Cosmology* (New York: Columbia University Press, 1960), 166–83.

10. For the full definition and justification of the Zero Principle, showing its derivation from Anaximander's relational ontology, see section 4.3.1.

CHAPTER 2. THE ANAXIMANDRIAN COMPASS

Anaximander thus provides the **first relational ontology** in Western thought—a metaphysics of process, not permanence; of relation, not substance; of becoming, not being. He opened the space that Aristotle would later fill with substance-categories, but his original insight points toward the very logic of identity that NPN recovers and formalizes. For a formal etymological and historical defense of the Zero Principle's roots in Presocratic thought, **see Appendix C: Anaximander and the Zero Principle.**

2.4 Heraclitus: The *Logos* of *Physis*

The Insight: Heraclitus of Ephesus looked at *Physis*—**dynamic objective reality—and saw its fundamental character: perpetual flux.**[11] He famously declared that one cannot step into the same river twice.[12] But he was no champion of chaos. He saw that this ceaseless change was not random; it was governed by a universal principle inherent in *Physis* itself, which he called the *Logos* (Λόγος)—a hidden, divine reason that structures all Becoming.[13] He saw reality as a dynamic tension of opposites, a "strife" that is, in fact, a harmonious *Dikē*.[14]

The Synthesis: Heraclitus provides the first coherent vision of a reality defined by Diachronic Primacy. He was not merely describing the *Logos* as a driver *within Physis*; he was identifying the *Logos* **as the very being of *Physis*.**

---

11. For Heraclitus' conception of the *Logos* as the structural principle unifying flux and opposition, see Charles H. Kahn, *The Art and Thought of Heraclitus: An Edition of the Fragments with Translation and Commentary* (Cambridge: Cambridge University Press, 1979), 19–28, 87–105.

12. For the river fragments, see Heraclitus in Kirk, Raven, and Schofield, *The Presocratic Philosophers*, KRS 214. NPN interprets this not as "chaos," but as the definition of a *Process Ontology*: the entity's identity is maintained *by* the flow, not in spite of it.

13. For the Logos as the universal principle and measure structuring all things, see Heraclitus DK 22 B1, analyzed in Kirk, Raven, and Schofield, *The Presocratic Philosophers*, 194–96. This marks the introduction of *Intelligent Naturalism*: the 'logic' is internal to the system, not imposed by an external god.

14. For the tension of opposites and the "back-stretched connexion," see Heraclitus in Kirk, Raven, and Schofield, *The Presocratic Philosophers*, KRS 207–9. This anticipates the NPN *Polarian Dynamic* (FP3).

His central, revolutionary insight is that **the "stuff" and the "law" are a unified reality.** His primordial element, "Ever-Living Fire," is the perfect symbol of this: Fire is not an inert substance that *has* a pattern; it *is* a pattern of transformative process. It is *Hylē* and *Logos* fused into a single, dynamic actuality.[15]

The *Logos*, therefore, is not a separate controller but the intelligible structure of the cosmic process itself. The flux is not an attribute of being; being *is* the flux, and the *Logos* is its measure. This is why he states that "all things come to pass in accordance with this *Logos*." He provides the definitive description of a reality where Becoming is ontologically primary: *Physis* is a lawful process of perpetual change, and to understand it is to grasp its *Logos*.

2.5 Empedocles: The Engine of Complexity

The Insight: Empedocles attempted to save the phenomena of the changing world by proposing a pluralistic solution. Four eternal "roots" (Earth, Air, Fire, Water) are mixed and separated by **two cosmic forces: *Philia* (Φιλία)—Love, which unifies, and *Neikos* (Νεῖκος)—Strife, which divides.**[16]

The Synthesis: Empedocles identified the dual drivers of cosmic and biological evolution. *Philia* and *Neikos* are not merely emotions; they are fundamental physical principles. *Philia* is the force of attraction, combination, and symbiosis—the drive toward complex unity. *Neikos* is the force of repulsion, competition, and individuation—the necessary friction that tests and breaks down forms. This is the Presocratic precursor to the

---

15. For Fire as the fundamental cosmological principle, see Heraclitus in Kirk, Raven, and Schofield, *The Presocratic Philosophers*, 197–200. Fire represents the *unity of substance and process*—it only exists while consuming and transforming. It is the archetype of the NPN *Archē*.

16. For Empedocles' four roots and the two forces of Love (*Philia*) and Strife (*Neikos*), see Kirk, Raven, and Schofield, *The Presocratic Philosophers*, KRS 346 (roots) and KRS 360 (forces). NPN secularizes these as non-anthropomorphic kinetic drivers: Attraction/Binding and Repulsion/Differentiation.

CHAPTER 2. THE ANAXIMANDRIAN COMPASS

evolutionary algorithm of variation and selection, the Yin and Yang of a creative universe.[17]

2.6 The Atomists: The Hardware of the Cosmos

The Insight: Leucippus and Democritus offered the most economical solution to the Parmenidean challenge. They proposed **two realities: the Full (atoms) and the Empty (the void).** Atoms—eternal, indivisible particles—move through the void, and their countless combinations constitute the pluralistic world of our experience. Everything, including the *Psyche*, is atoms.[18]

The Synthesis: The Atomists provided the material blueprint, a radical reduction of the *Archē* to its constituent "stuff" (*Hylē*). They perfectly described a deterministic, mechanical level of reality—the hardware of the cosmos (atoms and void). However, their system leads inexorably to a hard determinism, dismissing consciousness and will as illusory byproducts. They provided a powerful map of the *Hylē*, but their *Hylē*-only model failed to account for the reality of the emergent *Nous*—the navigator that uses the map. Theirs is a universe of being, with no coherent account of becoming or the knower.[19]

2.7 Parmenides: The Architecture of the Subject

The Insight: Parmenides of Elea, using pure logic, made a shocking claim: Change, motion, and plurality are illusions. True reality, "What Is," is one,

---

17. The interpretation of Empedocles' *Philia* and *Neikos* as proto-evolutionary forces finds a modern analogue in Richard Dawkins' gene-centered view of evolution. NPN formalizes this necessity as the "Entropic Mandate" (Theorem T5): without *Neikos*, the system cannot adapt and succumbs to entropy. See Richard Dawkins, *The Selfish Gene* (Oxford: Oxford University Press, 1976), 2–12, 88–108.

18. For the atomic theory of Leucippus and Democritus, see Kirk, Raven, and Schofield, *The Presocratic Philosophers*, KRS 549.

19. For a comprehensive account of the intellectual trajectory from Presocratic *physis* to later process philosophy, see W. K. C. Guthrie, *A History of Greek Philosophy, Vol. 1* (Cambridge: Cambridge University Press, 1962–1969), 55–60. NPN integrates the Atomist material substrate but rejects their eliminative materialism regarding the *Nous*.

eternal, unchanging, and indivisible.[20] He created a stark rift between the world of reason (Truth) and the world of the senses (Opinion).

The Synthesis: The standard reading sees Parmenides as a brilliant dead end. Our synthesis reveals a different truth. Parmenides was not accurately describing the cosmos; **he was uncovering the innate logical structure of the subjective self.**[21] The unity, identity, and persistence he ascribed to "Being" are the very properties of a coherent consciousness. He discovered the "One" not as the all-encompassing reality, but as the internal architecture of the mind—the stable center from which we experience the Heraclitean "Many."

2.8 Xenophanes of Colophon: The Limits of the *Nous*

The Insight: Xenophanes saw that human knowledge is not a state of certain possession but a process of asymptotic approximation. He formulated this in two foundational fragments. First, he established a constitutive limit: "No man knows, or ever will know, the clear truth (*to saphes*) about the gods and about all the things I speak of; for even if one should chance to say what is exactly the case, nevertheless he himself does not know it; but opinion (*Doxa*) is fashioned over all things."[22] Here, *Doxa* (Δόξα)—opinion or seeming—is presented as the fundamental epistemic condition of mortals. He points to a domain that cannot be known with certainty; this is an early description of the *Apeiron*—that which, by its nature, lies beyond the possibility of empirical verification. Second, he provided the method for operating within this condition: "The gods did not

---

20. For Parmenides' doctrine of "What Is," see Kirk, Raven, and Schofield, *The Presocratic Philosophers*, KRS 296.

21. Standard interpretations (e.g., KRS 241–62) read this as a denial of the physical world. NPN diverges by reframing Parmenides' "Being" as the internal architecture of the *Nous*—the stable, unified subject required to navigate the Heraclitean flux of the external world.

22. Xenophanes, DK 21 B34; KRS 186. This establishes the NPN distinction between *Epistēmē* (justified model) and *Saphes* (absolute truth). The latter is impossible for a finite *Nous*.

reveal all things to mortals from the beginning; but mortals, by long seeking (*zētountes*), discover what is better (*ameinon*)."[23] **Knowledge is therefore diachronic, iterative, and progressive—a *Becoming* achieved through seeking.**

In his theological fragments, Xenophanes describes a "one god" who "sees as a whole, thinks as a whole, hears as a whole" and "shakes all things by the thought of his mind."[24] This is a significant conceptual move: he is straining to describe the origin of the cosmos itself, and the only causal analogy available to him is that of an ordering intellect. While Anaximander posited the *Apeiron* as a generative but impersonal ground, Xenophanes—and later Anaxagoras—attributes *intellect* (*Nous*) to this ultimate source. They speculate that what lies beyond our observable reality is not just a boundless principle but a *mind* that brought order into being.

Relation to Heraclitus: Xenophanes and Heraclitus articulate complementary truths. Heraclitus described the objective *Logos* of a reality in perpetual flux.[25] Xenophanes described the subjective, epistemic corollary: the *knowing* of this reality is itself a perpetual, seeking process. One maps the flux of the world; the other maps the flux of understanding.

The Synthesis: Xenophanes made a critical advance by diagnosing the knower within a reality bounded by the *Apeiron*. He saw that knowledge itself is a *Becoming*: our best models are justified not by certainty but by being *better* than prior attempts, refined through "long seeking." However, his attempt to characterize the *Apeiron* itself as a divine intellect highlights a perennial tension: the finite *Nous*, a product of the *Archē*, instinctively projects its own nature onto the ultimate source. This speculative overreach does not diminish his core epistemological achievement but clarifies the boundary he helped define. He thus laid the precise epistemic groundwork for the Socratic *Elenchus*: if our grasp of reality is inherently

---

23. Xenophanes, DK 21 B18; KRS 185. This validates the "Diachronic Turn" in epistemology: Truth is a function of time and iterative seeking, not immediate revelation. For the modern rehabilitation of Xenophanes as the first 'Critical Rationalist,' see Karl Popper, Conjectures and Refutations (London: Routledge, 1963), 152–53.
24. Xenophanes, DK 21 B23–B26; KRS 169–71.
25. Heraclitus, DK 22 B1; KRS 194.

Doxa—prone to projection and error, as his critique of anthropomorphic religion shows[26]—then philosophy's first task must be the purification of the knower itself, beginning with a rigorous critique of just such unwarranted projections.

Within the NPN framework, this projection is a category error. It attempts to model the *Apeiron*—the very domain defined by the absence of empirical data—using the only causal agent we know from within the *Archē*, namely, our own conscious intelligence. It is an understandable but unwarranted inference, revealing the mind's compulsion to model even the unmodelable.

2.9 The Unified Model: *Physis* and the Emergent *Nous*

From this collaborative reconstruction, a powerful synthetic model of *Physis* emerges, one that resolves the central tensions the Presocratics themselves identified.

The model rests on two pillars, derived from our architects:

- **The Contrast-Field of the *Apeiron*:** This is Anaximander's fundamental insight, formalized in the Zero Principle. The *Apeiron* is not merely an "unknown" or an "epistemic limit"—it is the **indeterminate complement required for any determinate system to exist**. It is the boundless background against which the *Archē* becomes a determinate cosmos, and against which the *Nous* becomes a determinate knower.

- **The Framework of *Physis*:** This is the Heraclitean reality—a dynamic, lawful process. It is the *Archē* understood as **Hylē-as-structured-by-*Logos***, whose constitutive dynamics are the Empedoclean interplay of **Philia and Neikos**. This is the objective, causal cosmos.

---

26. Xenophanes, DK 21 B15–B16; KRS 168. NPN uses this critique to dismantle all anthropomorphic projections onto the *Archē*, establishing the "Silent Apeiron" of the Metaphysics.

CHAPTER 2. THE ANAXIMANDRIAN COMPASS

- **The Emergent *Nous*:** This is the Parmenidean structure—but correctly identified not as the cosmos, but as the unified, self-identical, persistent point of view *within* the cosmos. It is not a ghost in the machine, but a **functional, emergent property** of the complex system we call a brain. It is the framework's most sophisticated product: a navigator evolved to model the *Logos* in order to ensure survival.

Xenophanes' Critical Corollary: Between these pillars lies the epistemic condition articulated by Xenophanes. He saw that the *Nous*, while capable of modeling the *Logos*, operates under a constitutive limit: it can never know the *Apeiron*—the boundless ground of the *Archē*—with certainty. All human knowledge is therefore *Doxa*, a provisional opinion refined through "long seeking." This insight completes the map: the *Nous* is not only emergent and functional but also **intrinsically fallible and finite**.

The central synthesis is this: The subjective *Nous* is not the foundation of reality, but its most intricate product. It is a localized, temporary state of the objective *Physis*, evolved to effectively interact with the *Logos* from which it arose, yet forever barred from certain knowledge of its own ultimate source.

This model elegantly resolves the ancient problems:

- **The One/Many Problem:** They are not contradictory. The "One" (subjective consciousness) is a high-level state *of* the "Many" (the objective, dynamic framework of *Physis*).
- **The Being/Becoming Problem:** Being (the stable subject) is a persistent pattern within the process of Becoming (the cosmic *Physis*). It is a standing wave in a flowing river.
- **The Mind/Matter Problem:** The mind is not a separate substance. It is what matter *does* when organized by the *Logos* into a system of sufficient complexity for navigation. The *Nous* is *Physis* become self-aware.

2.10 Conclusion: The Compass Restored

The journey through the first *physiologoi* is not an academic exercise. It is the recovery of a lost compass. Thales gave us the imperative to seek a unified *Archē*. Anaximander revealed the boundless ground (*Apeiron*) of that source. Heraclitus mapped the dynamic, lawful *Logos* of *Physis*. Parmenides charted the internal structure of the knower itself. Empedocles uncovered the engine of complexity—*Philia* and *Neikos*—that drives the system forward.

Xenophanes provided the critical lens through which to view this compass. He showed that the *Nous* which holds the compass is itself a limited instrument. His fragments establish that knowledge is a *Becoming*—a gradient of confidence built through seeking, not a state of certain possession. He thus framed the ultimate philosophical problem: how can a fallible navigator, prone to error and projection (*Doxa*), reliably use a map of a reality it can never fully grasp?

The synthesis of these insights—the model of the emergent, fallible *Nous* within the objective *Physis*—provides a coherent, naturalistic, and robust foundation for understanding our place in the universe.[27] It grounds us, not in skepticism or illusion, but in a reality that is both lawfully objective and meaningfully experienceable.

We do not end this chapter with a dismantled world, but with a restored one. The conversation with the *physiologoi* is not over; it has just been renewed. We now possess their compass—a map of *Physis* itself, and an understanding of the navigator who must read it. But a compass is useless in the hands of one who does not know how to hold it. A map is inert until a navigator learns to align its symbols with the territory.

Thus, the great project faced a new, more intimate crisis. If the *Nous* is the emergent navigator, yet clouded by *Doxa* and prone to project its

---

27. This emergent, naturalistic foundation for ethics resonates with Martha Nussbaum's "therapy of desire," which seeks to align human striving with a rationally understood nature. See Martha C. Nussbaum, *The Therapy of Desire: Theory and Practice in Hellenistic Ethics* (Princeton, NJ: Princeton University Press, 1994), 3–16.

CHAPTER 2. THE ANAXIMANDRIAN COMPASS 21

own image onto the *Apeiron*, what happens when that navigator's own instrument—its faculty of reason—is corrupted by false certainty? The compass of the *Logos* cannot be read by a mind that refuses to look, or that mistakes its own reflection for the territory.

It was at this precipice that a new figure emerged, turning the gaze of philosophy from the stars to the soul. He understood that the journey to apprehend the cosmic *Logos* must begin with the medical treatment of the *logos* within one's own *Psyche*. His method would be the *Elenchus*—to purge *Doxa* and achieve the fertile *Aporia* Xenophanes described.[28] His name was Socrates, and he was the physician the *physiologoi* never knew they needed.[29]

2.11 The Navigator's Focus: Terms for the Socratic Turn

The physician of the *Psyche* requires a specialized diagnostic toolkit. As we examine the therapeutic project of Socrates, the terms below define the crucial states of cognitive disease (Doxa), the instruments of philosophical treatment (Elenchus, Lysis), and the healthy, purified result (Aporia, Dikaiosynē).

---

28. Xenophanes, DK 21 B34 and B18. See James H. Lesher, *Xenophanes of Colophon: Fragments* (Toronto: University of Toronto Press, 1992), 150–69, regarding the distinction between certain truth and the opinion fashioned over all things.

29. On philosophy as a therapeutic way of life, with Socrates as its archetypal practitioner, see Pierre Hadot, *Philosophy as a Way of Life: Spiritual Exercises from Socrates to Foucault*, trans. Michael Chase (Oxford: Blackwell, 1995), 81–125.

Table 2.1   The Toolbox of the Physician: Key Terms for Chapter 3

NPN Term	Role	Technical Definition
Psyche (Ψυχή)	The Soul	The stratified, emergent structure of the human mind; the evolved layering of the core *Hormē* into the fourfold system of *Nous, Logistikon, Thymos,* and *Orexis*.
Doxa (Δόξα)	Sickness	Unexamined belief or opinion; the cognitive misalignment rooted in contradictory models that the *Elenchus* seeks to cure.
Elenchus (Ἔλεγχος)	Diagnostic	The rigorous, logical cross-examination applied to break down *Doxa* and expose internal contradiction.
Lysis (Λύσις)	Dissolution	The destructive, necessary first stage of cognitive transformation; the breaking down of a rigid, flawed cognitive structure.
Katharsis (Κάθαρσις)	Purification	The conscious cleansing of the *Psyche* that follows *Lysis*, leaving the mind in a purified, sterile state.
Aporia (Ἀπορία)	Start Point	The productive state of intellectual void; the conscious recognition of a flawed model and the essential starting point for genuine inquiry.
Dikē (Δίκη)	Justice	The impersonal, causal justice of the *Archē*; the necessary constraints enforced by the *Archē* itself.
Dikaiosynē (Δικαιοσύνη)	Harmony	The functional harmony of the *Psyche*, where the *Logistikon* rightly governs *Orexis* and *Thymos* under the direction of the *Nous*.
Daimonion (Δαιμόνιον)	Inner Voice	The intuitive, guiding faculty of the *Nous*; the internal compass that signals misalignment and prompts course-correction.
Aretē (Ἀρετή)	Excellence	The optimal functioning of a thing according to its nature; in the *Psyche*, the full realization of *Dikaiosynē*.

Chapter 3

The Last Pre-Platonic: Socrates as the Physician of the Psyche

> "Although this account holds forever, people prove unable to understand it, both before they have heard it and when once they have heard it." — Heraclitus[1]

> "For I do nothing but go about persuading you all, old and young alike, not to take thought for your persons or your properties, but first and chiefly to care about the greatest improvement of the soul." — Socrates[2]

3.1 Introduction: The Misplaced Philosopher

The standard taxonomy of "Presocratic philosophy" has long provided a convenient chronological marker for the first thinkers of the Western tradition. Yet this classification, by defining an entire intellectual movement solely by its position before Socrates, creates a profound philosophical misrepresentation. It frames these thinkers as a mere prelude—primitive

---

1. Heraclitus, DK 22 B1. This fragment identifies the cognitive pathology Socrates addresses: the inherent difficulty of the *Nous* to grasp the *Logos* without therapeutic intervention. In Kirk, Raven, and Schofield, *The Presocratic Philosophers*, 187.

2. Plato, *Apology*, 30a–b. NPN interprets "care of the soul" not as religious piety, but as the maintenance of the *Psyche's* functional alignment with reality. In *Plato: Complete Works*, ed. John M. Cooper (Indianapolis, IN: Hackett Publishing Company, 1997).

stumbles toward the mature philosophy of the Athenian triad.³ This chapter proposes a corrective category: the **"Pre-Platonic" thinkers**. This epoch is defined not by chronology, but by a shared commitment to a specific mode of inquiry that sought the fundamental principles of reality—*Physis* (Φύσις)—without recourse to mythological explanation.

Within this framework, Socrates emerges not as the founder of a new tradition, but as the culmination and ultimate expression of the Pre-Platonic spirit. His singular mission was to act as a physician of the *Psyche* (Ψυχή), diagnosing and curing its most fundamental ailment: the sickness of false knowledge.⁴ His therapeutic method aimed to induce a state of intellectual *Katharsis* (κάθαρσις)—purification—through a process of philosophical *Lysis* (λύσις)—release or dissolution, which was the essential starting point for all genuine understanding.⁵

3.2 The Pre-Platonic Project: The *Logos* of *Physis* as the Standard of Health

The Pre-Platonic thinkers were united by a revolutionary project: to explain the cosmos through rational inquiry. Their collective goal was to understand *Physis*—the essential nature and origin of all things⁶—by identifying the *Archē* (Ἀρχή), the fundamental reality constituted by *Hylē* (Ὕλη) as structured and ordered by the *Logos* (Λόγος). This was a decisive turn toward a universe governed by intelligible, impersonal law.

---

3. For the canonical narrative of Presocratics as precursors, see W. K. C. Guthrie, *A History of Greek Philosophy, Vol. 1* (Cambridge: Cambridge University Press, 1962), 1–30; and Patricia Curd, "Presocratic Philosophy," in *The Stanford Encyclopedia of Philosophy* (2012). NPN rejects this teleological historiography. The Presocratics are not 'proto-Platonists' but independent *Physiologoi* whose project Socrates completes, not abandons.

4. For Socrates' description of his philosophical mission as a service to the god (therapist of the *Logos*), see Plato, *Apology*, 29d–31c. On the necessity of intellectual dissolution (*lysis*) before knowledge can be acquired, cf. Plato, *Meno*, 80a–b.

5. This therapeutic conception of Socratic philosophy aligns with Martha Nussbaum's analysis of Hellenistic ethics as a "therapy of desire" aimed at curing false beliefs about value. See Martha C. Nussbaum, *The Therapy of Desire: Theory and Practice in Hellenistic Ethics* (Princeton, NJ: Princeton University Press, 1994), 13–47.

6. Kirk, Raven, and Schofield, *The Presocratic Philosophers*, 1–10, 72–74.

In their search, they forged a powerful lexicon. Anaximander described a cosmic justice—*Dikē* (Δίκη)—inherent in the *Apeiron* (Ἄπειρον), where things "make reparation to one another for their injustice according to the assessment of Time."[7] This is more than a poetic image of balance; it is a profound articulation of a dynamic, lawful framework. The 'injustice' is the temporary dominance of any one thing or quality (e.g., Hot over Cold, Summer over Winter). The 'reparation' is the necessary rebalancing enforced by the framework itself. In this, Anaximander lays the ontological groundwork for the opposing forces—later articulated by Empedocles as *Philia* (Φιλία)—Love or Attraction—and *Neikos* (Νεῖκος)—Strife or Repulsion[8]—whose perpetual tension creates the cycles of the cosmos and the conditions for all adaptation and change. Heraclitus would later identify the *Logos* itself as the universal reason underlying this perpetual flux.[9] For these thinkers, to be in accordance with the *Logos* was to be in a state of cosmic "health" or balance. A thing's excellence—*Aretē* (Ἀρετή)—was its capacity to fulfill its function within this lawful whole.

The project remained, however, cosmic in scale. They provided the standard of health for the universe, but the diagnostic procedure for the individual human *Psyche*—the part of the universe that seeks to know—was still undeveloped.

3.3 Socrates the Physician: The Diagnosis of Universal Cognitive Sickness

Socrates executed a pivotal turn in the Pre-Platonic project. Where his predecessors sought to diagnose the cosmos, Socrates turned his attention to the human *Psyche* (Ψυχή). His practice revealed a near-universal condition: a cognitive sickness rooted in unexamined, contradictory, and thus misaligned belief—*Doxa* (Δόξα).

---

[7]. For Anaximander in Kirk, Raven, and Schofield, *The Presocratic Philosophers*, KRS 110. This formulation anticipates the NPN *Zero Principle*: justice is the necessary ontological repayment of borrowed time and matter required for individuation.

[8]. For Empedocles' two cosmic forces, see Kirk, Raven, and Schofield, *The Presocratic Philosophers*, KRS 360. NPN identifies these as the 'Kinetic Drivers' of the *Archē*, secularizing them as Attraction and Repulsion.

[9]. Kirk, Raven, and Schofield, *The Presocratic Philosophers*, 194–96.

He diagnosed a ubiquitous ailment of the *Psyche*, a state where its innate *Hormē* (Ὁρμή)—its impulse to strive and persist—was frustrated and misdirected by a flawed understanding of the *Logos*. His mission was to practice a form of philosophical triage, identifying the severity of this misalignment in every *Psyche* he encountered.

His primary diagnostic tool was the *Elenchus* (Ἔλεγχος)—cross-examination. This was not a gentle conversation but a rigorous examination, akin to a physician's probing of a wound.[10] Its purpose was to test the internal coherence of a person's understanding of fundamental concepts—Justice, Piety, Courage—inevitably revealing the symptoms of the soul's disease: contradiction, inconsistency, and confusion. No one passed the examination.[11]

Complementing this public diagnosis was his private, inhibitory *Daimonion*. This 'divine sign' can be understood as Socrates' own highly attuned faculty of instinctual logic—a hardwired, pre-rational attunement to the *Logos*. It was not a source of positive knowledge but a negative guide, a sudden intuition of misalignment with the coherent structure of reality, preventing him from committing a logical or ethical error.[12] It was his own internal diagnostic safeguard, ensuring his method remained aligned with the very *Logos* he sought to help others apprehend.

Having diagnosed the disease of the soul, Socrates' method aimed at a specific, two-stage cure.

---

10. For Gregory Vlastos's influential analysis of the Socratic elenchus as both a logical refutation and a therapeutic tool for exposing moral inconsistency, see Gregory Vlastos, *Socrates: Ironist and Moral Philosopher* (Cambridge: Cambridge University Press, 1991), 107–31.

11. For a prime example of the Elenchus exposing contradiction and inducing aporia, see Plato, *Euthyphro*, 11a–b. The dialogue demonstrates that without *Elenchus*, the subject remains unaware of their own cognitive dissonance.

12. For Socrates' description of his *daimonion*, see Plato, *Apology*, 31c–d. NPN naturalizes the *Daimonion* not as a spirit, but as the *Somatic logos*—the high-speed, intuitive processing of the *Thymos* alerting the *Nous* to danger (see Section 6.3, 'Topography of Navigation').

3.4 The Therapeutic Goal: *Lysis* and *Katharsis* as the Path to *Aporia*

The goal of Socratic therapy was not to impart new dogma but to cure the *Psyche* of its existing sickness. This cure was a two-stage process:

- **Lysis (Λύσις)**—a releasing, loosening, or dissolution.[13] This was the destructive action of the *Elenchus*. Socrates sought to dissolve the rigid, calcified structures of false belief that imprisoned the mind. He applied the solvent of logical scrutiny until the interlocutor's certainty began to break down. The dialogues of *Euthyphro* and *Meno* are perfect records of this process of *Lysis*, where confident knowledge is systematically dissolved.[14]

- **Katharsis (Κάθαρσις)**—a conscious purification or cleansing. This was the intended result of the *Lysis*. Once the false beliefs were dissolved and released, the *Psyche* was left in a state of purified emptiness—the state of *Aporia*. This *Aporia* is not a failure but the successful outcome of the treatment. It is the "clean wound" after the removal of a foreign object or infected tissue. It is the intellectual void, free from the pollution of unexamined *Doxa*, now ready to receive genuine knowledge—*Epistēmē* (Επιστήμη).

The famous Socratic declaration, "I know that I know nothing," is the epitome of this purified state.[15] It is not ignorance, but the pronouncement of a *Psyche* that has undergone *Katharsis* and is now intellectually sterile, healthy, and prepared for truth. It is the state of a *Psyche* whose *Hormē* is no longer shackled by *Doxa* and is free to seek genuine alignment with the *Archē*.

---

13. This process of cognitive and emotional *lysis* (dissolution) as a precondition for philosophical transformation corresponds to Pierre Hadot's concept of "spiritual exercises." See Pierre Hadot, *Philosophy as a Way of Life: Spiritual Exercises from Socrates to Foucault*, trans. Michael Chase (Oxford: Blackwell, 1995), 81–90.

14. For the process of *Lysis* in action, see the entirety of Plato's *Euthyphro* and the first part of the *Meno* (up to 80a). These are not failed conversations, but successful surgeries.

15. See Plato, *Apology*, 21d. NPN redefines Socratic Ignorance as 'Epistemic Sterility'—the necessary removal of *Doxa*-pathogens to allow for the growth of *Epistēmē*.

3.5 Socrates as the Pre-Platonic Culmination: Applying the Cosmic Standard to the Psyche

Seen through this therapeutic lens, Socrates is the culmination of the Pre-Platonic project. The thinkers who preceded him—from Anaximander to Heraclitus—forged the concept of a cosmos governed by a lawful *Logos*, the standard of cosmic *Aretē* (excellence). Socrates took this cosmic standard and applied it directly to the microcosm of the human *Psyche*.

He does not provide the answers about *Physis*; he provides the cure for the cognitive ailments that prevent us from finding them. He is the physician who prepares the patient for health by purifying the soul's *Hormē*, clearing the blockages so it can flow in accordance with the *Logos*. The Pre-Platonics described cosmic *Aretē*; Socrates developed the therapeutic practice to achieve psychic *Aretē*. He is, therefore, the last and most necessary of the Pre-Platonics, the one who completed their project by developing its essential clinical methodology.[16]

3.6 The Platonic Divergence: From Therapy to Metaphysical Prosthesis

The steadfastness with which Socrates resided in the therapeutic state of *Aporia* marks the end of this epoch; his student, Plato, represents the decisive divergence.[17] Plato could not abide the purified emptiness that was the result of Socratic *Katharsis*. In response, he constructed a positive, comprehensive system—the Theory of Forms—as a kind of metaphysical prosthesis for the soul. Where Socrates sought to cure the *Psyche* of its

---

16. For the culmination of Presocratic rational inquiry, see Jonathan Barnes, *The Presocratic Philosophers* (London: Routledge and Kegan Paul, 1982), 467–77. Barnes emphasizes that the Presocratics established a rigorous, rational methodology; Socrates applies this "Milesian Method" to the human sphere, concluding the inquiry into *physis* by locating the instrument of inquiry (*Nous*).

17. On Plato's systematic departure from the Socratic therapeutic model and the naturalistic foundation of the Presocratics toward a positive metaphysical system, see Julia Annas, *Platonic Ethics, Old and New* (Ithaca, NY: Cornell University Press, 1999). For the general ethical divergence, see 1–27; for the specific shift from purgative inquiry to constructive system-building, see 52–76.

need for false certitude, Plato provided it with a new, grander certitude to cling to.[18]

This marks the true schism: the end of the therapeutic, purgative inquiry of the Pre-Platonics and the beginning of the constructive, system-building era of Platonic philosophy.

3.7 Conclusion: The Unexamined Soul is Not Worth Healing

The "Pre-Platonic" epoch is defined by a shared commitment to a rigorous, first-principles inquiry into the nature of a law-governed reality. Socrates stands as its final and most profound exponent, demonstrating that the journey to understand the *Logos* of the cosmos must begin with the medical treatment of the *logos* within one's own soul. His legacy is not a doctrine but a permanent therapeutic practice:[19] the diagnosis of false belief, the *Lysis* of rigid dogma, and the pursuit of intellectual *Katharsis* as the only healthy ground for understanding. He bequeathed to us the unsettling and vital truth that the unexamined soul is not merely unworthy, but unwell.

The Socratic practice, however, culminates in a purified void—a state of productive *Aporia*. It is from within this Socratic void that we now turn to the systematic task of presenting the NPN meta-structure itself, built with the very tools our predecessors first forged.

---

18. For the critique of *anamnesis* and the premature rationalism of the Platonic response to Socratic skepticism, see Vlastos, *Socrates: Ironist and Moral Philosopher*, 45–67. NPN argues that *anamnesis* functions as a 'cheat code' to bypass the existential difficulty of *Aporia*.
19. Nussbaum, *The Therapy of Desire*, 13–47.

Chapter 4

The Metaphysical Structure of Neo-Pre-Platonic Naturalism

"I shall tell you a twofold truth. At one time one alone grew from many, and at another many grew from one alone: double is the birth of mortal things, and double their failing." – Empedocles[1]

4.1 Introduction:
The Architecture of a First-Principles System

The collaborative reconstruction of the Presocratic blueprint in the previous chapters provides the raw materials. This chapter undertakes the systematic task of assembly, presenting the consolidated meta-structure of Neo-Pre-Platonic Naturalism. What follows is a unified framework for understanding reality, consciousness, and knowledge, synthesized from their foundational insights and crystallized into a set of formal first principles. This entire structure is unified by a decisive methodological commitment: the **Diachronic Turn**.[2]

---

    1. Empedocles, Fragment 17 (KRS 348). This fragment establishes the "Diachronic Heartbeat" of the NPN system: reality is not a static object but a reciprocating engine driven by the exhaustive polarity of *Philia* and *Neikos*. In G. S. Kirk, J. E. Raven, and M. Schofield, *The Presocratic Philosophers: A Critical History with a Selection of Texts*, 2nd ed. (Cambridge: Cambridge University Press, 1983), 287.

    2. This emphasis on first principles as the ground of a coherent system aligns with Alfred North Whitehead's process metaphysics. NPN adopts Whitehead's primacy of "becoming" but rejects his theological speculations, grounding the process solely in the naturalistic *Archē*. See Alfred North Whitehead, *Process and Reality: An Essay in Cosmology*, corrected ed. (New York: Free Press, 1978), 3–18.

This turn is not a mere preference but a necessary correction demanded by a fundamental truth. To understand any entity—from a particle to a person to a polity—is to understand its history and trajectory. A static analysis can only provide a frozen cross-section, a map of being that ignores the reality of Becoming. The *Physis* is not a static arrangement but a dynamic, evolutionary process. The *Nous* is not a pre-programmed substance but an emergent navigator. Knowledge is not a state of certain possession but a process of continuous model-revision. The entire meta-structure that follows is an exposition of this single, inescapable principle.[3]

> **First Principle**
>
> **FP2: FIRST PRINCIPLE OF BECOMING: DIACHRONIC PRIMACY**
>
> **Statement:** Being is a stabilized pattern within Becoming. A synchronic state is a derived abstraction; **Becoming** is ontologically primary.
>
> **Justification:** The negation of this principle posits a world without change—a static, timeless snapshot as fundamental reality. This is not merely false but logically incoherent and a performative contradiction.

This is a completely new system, and learning it will be challenging—but the key is to remember that we are simply restoring the missing dimension of time. A dimension that was systematically excluded from Western philosophy after Parmenides. By doing so, we resolve philosophy's seemingly unsolvable problems into a coherent whole, unifying previously separate schools of thought in the process.

---

3. The crisis of epistemic certainty that follows this schism was first anticipated by the Presocratics, particularly Xenophanes, who established the crucial distinction between fallible human opinion Doxa (Δόξα) and objective truth Aletheia (Ἀλήθεια). See James H. Lesher, *Xenophanes of Colophon: Fragments* (Toronto: University of Toronto Press, 1992), 150–65, esp. fragments B34.

It is a demanding intellectual shift—from inspecting static snapshots to navigating dynamic reality—but one that is ultimately rewarded with a complete understanding of the *Physis* itself.[4]

4.2 First Philosophy: The Geometry of Distinction

Before we articulate the First Principles, we must address a more fundamental question: **By what necessity do these principles exist?** Are they merely observations of a specific universe, or are they the structural requirements of *any* possible reality?

To answer this, we must descend to the level of **First Philosophy**—the study of the conditions of possibility for existence itself.

In **Appendix K: First Philosophy: The Boundary Condition**, we provide a formal geometric proof of the system that follows. By analyzing the simplest possible cognitive and ontological act—the drawing of a **Distinction** (a boundary)—we derive the necessity of the entire NPN framework.

This derivation reveals that the structure of reality is not arbitrary. If a thing exists as a determinate entity, it must satisfy the **Boundary Condition**:

1. **The Necessity of the Gap:** To be distinct is to be bounded; to be bounded is to be separated from others; therefore, an interstitial space must exist.
2. **The Necessity of the Apeiron:** This interstitial space cannot itself be a bounded entity (lest we trigger an infinite regress). It must be the *Indeterminate Ground* (The *Apeiron*).
3. **The Necessity of Navigation:** To exist in such a field is to be required to maintain one's boundary against the indeterminate. Thus, Ethics is not a social invention but a geometric imperative.

---

[4]. The fundamental methodological move of prioritizing process over static substance aligns with Process Philosophy. For a modern systematic defense of process ontology that parallels NPN's rejection of static substance, see Nicholas Rescher, *Process Metaphysics: An Introduction to Process Philosophy* (Albany, NY: State University of New York Press, 1996), 1–15.

While the chapters that follow describe the *physics* and *mechanics* of the Navigator's world, Appendix K describes its *geometry*. It proves that the "War of the Giants" over the nature of Being was not a clash of opinions, but a failure to recognize the logic of the Boundary.

The reader is invited to review **Appendix K** for the full rigorous derivation. What follows here is the operational system that emerges from those geometric roots.

4.3 The Ground: The General Zero Principle and Its Application

Before we can identify the fundamental bedrock of the *Archē*, we must acknowledge the transcendental condition that allows any identity to exist. Existence is not a brute fact occurring in isolation; it is a relational emergence.

Consider a simple mental exercise: **"Think of a dog."** When you visualize the dog, your attention is fixed on a determinate form—the figure of the animal. However, the dog only possesses a distinct identity because it exists against a "black background"—the field of contrast that is not-the-dog. Without that background, the dog has no boundary; without a boundary, it has no form; and without form, it cannot be distinguished as a "dog".

This logic applies to the total scale of **generality**. It is the "black background" of contrast that gives anything its form. For any system to be determinate, it must emerge from an indeterminate field. What follows are the formal principles that articulate this foundational insight, beginning with its most general form.

4.3.1 The General Zero Principle: The Escape from Infinite Regress

Every philosophical system faces the challenge of grounding determination without falling into infinite regress. If A is defined by not being B, and B by not being C, we enter an endless chain. The only escape is to acknowledge that the ultimate foundation cannot itself be determinate. The same is true for circular logic: A is defined as not-B, and B is defined as not-A. This is not a solution but a special case of infinite regress—a closed loop that provides no more foundation than an open chain. Both infinite regress and circular reasoning fail to provide ultimate grounding because they never reach a stopping point that is not itself dependent.

> **General Zero Principle**
>
> **GZP: GENERAL ZERO PRINCIPLE**
>
> **Statement:** For anything to possess determinate identity, meaning, or existence, it must exist within a delimited context set against an **indeterminate background**. The ultimate foundation cannot itself be determinate, for then it would require further foundation. It must be **indeterminate**.
>
> **Justification:** Infinite regress is avoided only by positing a ground that is not a "thing" among things. To define "A" requires "not-A". If "not-A" is also determinate ("B"), the regress continues. It must stop at the indeterminate.

For the complete formal derivation of the General Zero Principle (GZP) and its application to system identity Zero Principle (ZP), including a detailed analysis of its roots in Anaximander's Apeiron, Spinoza's negatio, and Spencer-Brown's calculus of distinction, see Appendix A.

4.3.2 The Specific Case: System Identity and the Zero Principle

From GZP follows a specific corollary that addresses the ontological condition for *system identity*—the formalization of what we observed with the mental image of the dog.

> **Zero Principle**
>
> **ZP: ZERO PRINCIPLE: THE NECESSITY OF CONTRAST**
>
> **Statement:** For any determinate system to exist, there must be an indeterminate complement —a not-system. Identity is not intrinsic but relational, defined by emergence from a contrasting field.
>
> **Justification:** To define a thing is to distinguish it from what it is not. If the context of distinction were itself fully determinate, an infinite regress would render definition impossible. To say a thing has an identity is to say it emerges from a field of contrast —a ground that is not the thing. To claim there is a system is to claim there is an "inside" to the system, which is meaningless unless there is an "outside" the system.

4.3.3 The Cosmic Application: The *Apeironic Context*

With ZP established, we can now articulate its direct application to the fundamental structure of reality itself.

> **Corollary**
>
> **C1: PRIMARY COROLLARY: THE APEIRONIC CONTEXT**
>
> **Statement:** The *Apeiron* is the necessary, indeterminate field that provides the ontological contrast for the *Archē*. It is not merely an epistemic limit of knowledge, but the ontological background required for the *Archē* to possess determinate identity.
>
> **Justification:** By the **Zero Principle**, identity is differential, not intrinsic. The *Archē* (the totality of determinate reality) therefore cannot be defined in isolation; it requires a complement. The *Apeiron* serves as that absolute complement—the context-field of indeterminacy that allows the *Archē* to be something rather than nothing.

4.3.4 The Paradox of the Boundless: ZP and the Miletian Source

This principle addresses the silent assumption underlying all formal logic and reality: **Determinacy requires Indeterminacy**. It establishes that identity is a differential relationship with a boundary rather than a property of a self-sufficient substance. From this ground, the relationship between the *Archē* and its necessary complement, the *Apeiron*, is established as a functional necessity.

This resonates with Spinoza's dictum *Omnis determinatio est negatio* (Letter 50 to Jarig Jelles, 1674).[5] For Spinoza, to "determine" or define a thing is to limit it, which necessarily implies a negation of the infinite background. In this context, the *Archē* is only possible through the stipulative negation of the *Apeiron*, establishing the **Zero Principle** as the ground upon which any finite determination is constructed.

4.4 The Foundation: The Primacy of the *Archē*

Every coherent system requires a grounding—a non-negotiable bedrock upon which its claims rest. Neo-Pre-Platonic Naturalism asserts that there is only one logical candidate for this foundation: reality itself. By anchoring the system in the *Archē*, we bypass the need for transcendent justifications and stand on the firm ground of what is.

4.4.1 The *Archē* as a Dynamic Hylomorphic Totality

Neo-Pre-Platonic Naturalism posits the *Archē*—the sole fundamental existence: the closed, causal system of objective, physical reality.[6] The *Archē*

---

5. See Benedict de Spinoza, *Complete Works*, ed. Michael L. Morgan, trans. Samuel Shirley (Indianapolis: Hackett Publishing, 2002), 892. This establishes the historical precedent for the "Zero Principle": identity is not intrinsic but differential. A thing is defined only by what it excludes (negation).

6. The concept of an *Archē* as the foundational, generative source of all things finds one of its earliest and most sophisticated expressions in Anaximander's *Apeiron*. See Charles H. Kahn, *Anaximander and the Origins of Greek Cosmology* (New York: Columbia University Press, 1960), 166–83.

is a hylomorphic unity, but this must be understood dynamically.[7] It is constituted by two interdependent, co-primal principles:

- ***Hylē*** (Ὕλη)—not a passive substrate, but the dynamic, generative, and constraining potentiality of the cosmos. It is the "hardware" of existence, possessing a "grain"—innate tendencies and limits (constants, quantum fields) that shape what forms are possible. It is the Yin-like principle of receptivity and novel possibility.

- ***Logos*** (Λόγος)—not merely a set of laws, but the active, structuring principle of actualization. It is the "software" that actualizes potential into determinate forms and processes. It is the Yang-like principle of determination and order.

The *Archē* is not *Hylē* plus *Logos* as separate components, but rather ***Hylē-as-actualized-by-Logos***. They are two aspects of a single, dynamic reality: the lawful substance of the universe in perpetual, co-constitutive interaction.[8]

> **Analogue: The Architecture of the Computer:**
> Consider a computer system. The processor is not a neutral piece of silicon (*Hylē*) onto which logic (*Logos*) is superimposed; the logic is etched into the physical arrangement of the gates. Similarly, "software" is often mistaken for immaterial logic, yet it exists only as physical states—magnetic domains on a drive or voltage traps in memory. To "write code" is to physically rearrange the material substrate. Hardware and software are not two different substances (mind and matter); they are simply two rates of physical mutability—two different ways to discribe the same thing. The *Archē* exists in this

---

[7]. Cf. Kirk, Raven, and Schofield, *The Presocratic Philosophers*, 108–17. The *Archē* is a primordial substance, resisting fixed or static categorization. NPN formalizes this as "Process Substance"—matter defined by its rate of change.

[8]. The ordering of the First Principles in this chapter is **non-sequential**; it is designed to prioritize rhetorical flow and conceptual clarity, starting with foundation and building complexity, rather than adhering to the strict deductive order of the formal system established in **Appendix B**.

mode: the laws of physics are not imposed upon the cosmos from the outside; they are the inherent, structural behavior of the cosmos itself.

> **First Principle**
>
> **FP1: FIRST PRINCIPLE OF REALITY:
> THE PRIMACY OF THE *ARCHE***
>
> **Statement:** The *Archē* is. It is the fundamental, objective, physical reality that exists.
>
> **Justification:** To deny it is self-refuting: the very act of denial must be performed *within* a reality by a conscious entity that *exists*. Solipsism and radical skepticism are not arguments against the *Archē*; they are linguistic games played within it. The *Archē* is the non-negotiable ground of all inquiry.

4.4.2 The *Apeiron* (Ἄπειρον) as the Boundary of the *Archē*

The *Apeiron* is the **fundamentally and permanently unknowable** context for the *Archē*. It is the absolute limit of the *Logos*. Questions that seek to go "outside" or "before" the *Archē* (e.g., 'What caused the *Archē*?') are category errors that point to the *Apeiron*.[9] To speculate about it is to generate *Doxa* (Δόξα)—unexamined belief, not *Epistēmē* (Ἐπιστήμη)—justified knowledge.[10]

4.4.3 The Dynamic Interface: Navigating the Not-Yet-Known *Archē*

The process of inquiry occurs entirely within the *Archē*. The *Archē* itself contains two epistemic states for a knowing subject:

---

[9]. The *Apeiron* as an absolute epistemic and ontological limit parallels Nicholas Rescher's process-philosophical insistence on the inexhaustibility of reality and the finitude of human cognition. See Rescher, *Process Metaphysics*, 1–15, 92–108.

[10]. For the distinction between Truth and Opinion (*Doxa*), see Kirk, Raven, and Schofield, *The Presocratic Philosophers*, 179–80.

1. **The Known *Archē*:** Reality as it has been successfully modeled. We have a high-confidence grasp of the *Logos* governing a specific domain of *Hylē*.
2. **The Not-Yet-Known *Archē*:** All phenomena within the *Archē*—both undiscovered configurations of *Hylē* and unresolved aspects of the *Logos*—that are currently unmodeled but are, in principle, accessible to inquiry.

The diachronic process of discovery is the conversion of the "not-yet-known" *Archē* into the "known" *Archē*, a perpetual navigation bounded by the silent *Apeiron*.

> **First Principle**
>
> **FP8: FIRST PRINCIPLE OF RELATION: THE NAVIGABILITY OF THE *ARCHE***
>
> **Statement:** The *Logos* of the *Archē* is, in principle, model-able by an emergent subsystem within it. Reality is structured such that it can be successfully navigated.
>
> **Justification:** The negation is a performative contradiction. To claim "reality is fundamentally unnavigable" is to assert a model of reality (the "unnavigable" model) that you are, by the act of asserting it, successfully using to navigate. It is the claim, "My map of unmappability is correct." This is the deepest possible self-refutation.

4.5 The Boundary of Knowledge: The *Apeiron* and the Limits of Logic

The *Archē* is the domain of the knowable—the structured, lawful reality that can, in principle, be modeled. But every map has a boundary, and every system has an outside. For NPN, this absolute limit is the **Apeiron** (Άπειρον)—the boundless, indefinite context for the *Archē*. It represents the fundamental and permanent horizon of knowledge.

4.5.1 The *Apeiron* as the Unstructured Ground

The *Apeiron* is not merely "the unknown" in a provisional sense. It is the category for which **empirical observation is impossible**. It is the unstructured ground from which the determinate *Archē* emerges and to which it ultimately relates. Questions that seek to go "outside" or "before" the *Archē*—such as "What caused the Big Bang?" or "What exists beyond spacetime?"—are not difficult scientific questions; they are **category errors** that point toward the *Apeiron*. They are attempts to apply the categories of the *Logos* (like causality and substance) to a domain where those categories, by definition, do not apply.

4.5.2 The Impotence of Logic Before the *Apeiron*

This leads to a critical epistemological constraint. Logic—the *Somatic logos*—is a tool for processing empirical observations. It is the operating system for navigating the *Physis* (Φύσις). To apply pure logic to the *Apeiron* is to run a program with no input data. The output is not knowledge, but speculative nonsense—elegant, internally consistent, but utterly ungrounded *Doxa*.

First Principle

FP5: FIRST PRINCIPLE OF KNOWLEDGE: THE IMPOTENCE BEFORE THE *APEIRON*

Statement: The *Apeiron* is the category for which empirical observation is impossible. Therefore, no logical operation can be empirically grounded or validated.

Justification: Logic is the *Somatic logos*—a tool for processing empirical observations. To apply logic to a domain with zero observations is to run a program with no input data. The output is not knowledge, but nonsense. The negation is an attempt to generate information from an information vacuum, which is a logical absurdity.

Metaphysical speculations about the "ultimate nature" of reality that attempt to bypass empirical grounding are exercises in generating fiction. They are the *Nous* attempting to model the *Apeiron* using only its own internal architecture, a process that is doomed to produce castles in the sky. The *Apeiron* stands as the final check against rationalist hubris, a silent reminder that our most powerful cognitive tool has sharp and non-negotiable limits.

4.6 The Constitutive *Logos* (Λόγος): Lawful Structure in Nature

If the *Archē* is the fundamental substance of reality, then the *Logos* is its fundamental structure. Neo-Pre-Platonic Naturalism revives the Heraclitean conception of the *Logos* not as a divine word or a transcendental principle, but as the inherent, discoverable, and constitutive order of the natural world.

4.6.1 The Logos and the Zero Principle: Polarity as Necessity

The *Logos* provides the lawful grammar of the *Archē*, but the Zero Principle (ZP) establishes that this grammar must be fundamentally polar. Identity is not intrinsic but relational, defined by emergence from a contrasting field. For the *Logos* to actualize as a determinate system, it must manifest through the tension of opposing vectors.

4.6.2 The *Logos* as Generative Potential

The *Logos* is the fundamental, unified potential for all interaction and relation within the *Archē*. It is not merely a descriptive catalog of laws but the generative source from which all specific laws and forces manifest.

4.6.3 The Exhaustive Polarity: *Philia* and *Neikos*

For this potential for relation to become manifest in a non-uniform reality, it must actualize as a fundamental polarity. This polarity is exhaustive and necessary:

CHAPTER 4. NPN METAPHYSICS

- **Philia (Φιλία)** is the constitutive force of Attraction, Union, and Cohesion (the drive to move towards).
- **Neikos (Νεῖκος)** is the constitutive force of Repulsion, Division, and Selection (the drive to move away).[11]

A third fundamental mode of relation is metaphysically and conceptually impossible, as any conceivable interaction can only be defined as moving entities "toward" or "away from" one another. This duality is not a synchronic binary but a diachronic polarity—the fundamental vectors of the *Logos* in action.

4.6.4 Anaximander's Cosmic Justice: *The Impermanence of Form*

Anaximander's fragment—that things "make reparation to one another for their injustice according to the assessment of Time"—provides the first glimpse of the cosmic forces that would later be systematized as *Philia* and *Neikos*.[12] The emergence of any determinate entity is a temporary "injustice," a state of imbalance driven by the interplay of these forces. Its eventual dissolution is not a failure but a necessary rebalancing—a debt repaid to the whole through the impartial, diachronic metric of Time.

The **Zero Principle** establishes why *Philia* must have *Neikos* as its necessary contrasting background; identity is differential, and without this exhaustive polarity, the *Archē* could not possess determinate form. This fundamental principle demonstrates the need for the polarity by identifying that any determinate system requires an indeterminate complement to

---

11. For Empedocles' cosmic forces *Philia* (Love) and *Neikos* (Strife), see Kirk, Raven, and Schofield, *The Presocratic Philosophers*, 280–321. NPN interprets these not as mythical figures but as the necessary "Kinetic Drivers" of the *Archē*. Without Neikos, the *Archē* is a frozen block; without Philia, it is dispersed dust.

12. For Anaximander's concept of *Dikē* providing the ground for these forces, see G. S. Kirk, J. E. Raven, and M. Schofield, *The Presocratic Philosophers: A Critical History with a Selection of Texts*, 2nd ed. (Cambridge: Cambridge University Press, 1983), 110. For Empedocles' systematic articulation of the two cosmic forces of Love (*Philia*) and Strife (*Neikos*), see Kirk, Raven, and Schofield, *The Presocratic Philosophers*, 360. NPN unifies these accounts: Anaximander provides the *Temporal Metric* (reparation over time), while Empedocles provides the *Kinetic Mechanism* (Love/Strife).

be "something" rather than "nothing". By the Zero Principle, identity requires contrast; therefore, for a system of union (*Philia*) to be determinate, it requires the force of separation (*Neikos*) as its ontological background.

4.6.5 The Dual Engine of the Cosmos: *Philia* and *Neikos* Defined

> "At one time they [the elements] are all brought together by Love into one, at another each is borne apart by the hatred of Strife." — Empedocles[13]

This cycle is driven by two impersonal, amoral, and equally necessary forces, first systematically identified by Empedocles:[14]

- **Philia (Φιλία)** is the constitutive force of **Attraction, Union, and Cohesion**. In the language of the Zero Principle, *Philia* is the **Emergent Figure**—the "mark" made by gathering *Hylē* (Ὕλη) into a bounded, integrated identity. It is the principle of structured persistence and negentropy that allows a system to stand out as a distinct "Something" against the void.

- **Neikos (Νεῖκος)** is the constitutive force of **Repulsion, Division, and Selection**. In the language of the Zero Principle, *Neikos* provides the **Necessary Ground**—the "unmarked state" of separation and spacing that allows for plurality and distinctness. It is the principle of differentiation and the entropic background that eventually dissolves forms, returning them to the *Apeiron*.[15]

---

13. Empedocles, Fragment DK 31 B17. This fragment defines the "Diachronic Heartbeat" of the NPN system: reality is not a static object but a reciprocating engine driven by the exhaustive polarity of Philia and Neikos. These forces are amoral physical vectors—integrative and disintegrative—required for the "churn" of the cosmos.

14. Empedocles, Fragment 17 (KRS 348). In Kirk, Raven, and Schofield, *The Presocratic Philosophers*, 287. Empedocles is often misread as a dualist; NPN reads him as a functionalist. *Love* and *Strife* are not moral categories but physical vectors—Integrative and Disintegrative—required for the "churn" of the cosmos.

15. This structural requirement for a background of decay mirrors Erwin Schrödinger's observation that life must actively resist the "returning" to thermodynamic equilibrium—a state of maximum entropy—by extracting "negative entropy" (the work of *Philia*) from its environment. See Erwin Schrödinger, *What is Life? The Physical Aspect of the Living Cell* (Cambridge: Cambridge University Press, 1944), 72–75.

> **First Principle**
>
> ## FP3: FIRST PRINCIPLE OF COSMIC DYNAMICS: THE LOGOS AND ITS EXHAUSTIVE POLARITY
>
> **Statement:** The *Archē* satisfies the General Zero Principle through the fundamental polarity of *Philia* and *Neikos*. The *Logos* is the tension between the emergent figure of Order (*Philia*) and the necessary background of Separation (*Neikos*).
>
> **Justification:** This is the translation of the Zero Principle into physics:
>
> 1. **The Necessity of Ground (Neikos):** According to the GZP, a determinate thing cannot exist without a contrasting context. In the *Archē*, *Neikos* (Entropy/Separation) provides the "unmarked state" or distinctness required for any form to exist.
> 2. **The Emergence of Figure (Philia):** *Philia* is the "mark" of identity—the gathering of *Hylē* into a bounded form against the background of *Neikos*.
> 3. **Exhaustion:** There is no third state. A physical system is either integrating (becoming figure) or disintegrating (returning to ground).
> 4. **Ontological Closure:** Because identity is relational (ZP), the *Logos* must manifest as a polarity. A singular force would lack a contrasting background, collapsing the *Archē* into an undifferentiated, static unity where change—and thus the *Archē* itself—would cease to exist. Conversely, absolute division without the unifying force of *Philia* would prevent the formation of any persisting systems, leading to a total fragmentation into the *Apeiron* where identity is likewise impossible.

Table 4.1 The Core Drivers: *Philia* and *Neikos*

Feature	Philia (Φιλία) (The Drive to Union)	Neikos (Νεῖκος) (The Drive to Separation)
Core Principle	Attraction, Cohesion, Synthesis, Integration	Repulsion, Division, Separation, Individuation
Cosmic Manifestation	Gravity, Chemical Bonding, Nucleosynthesis	Entropy, Radioactive Decay, Diffusion
Biological Function	Inheritance (Preserves Shared Code), Symbiosis, Cell Cohesion	Variation (Mutation), Competition, Natural Selection (Filtering)
Role in *Nous*	Cognitive Coherence, Model Building, Trust, Loyalty	Doubt, Differentiation, Boundary Setting, Critical Analysis

4.6.6 A Unified View: The Dance of Forces in Evolution and Complexity

The interplay of **Philia** (Φιλία) and **Neikos** (Νεῖκος) is the engine of all complex phenomena. The ultimate demonstration of this dynamic polarity lies in the process of evolution, which provides a perfect diachronic model of their functions.[16]

Philia (Φιλία) operates as the **engine of inheritance and cohesion**. It is the force that binds lineages together, propagating shared genetic information through time. It manifests as the common blueprint uniting a species.

Neikos (Νεῖκος) operates with a dual function. It is first the **engine of variation and individuation**, driving divergence through mutation and competition. Its second, equally crucial role is as the **purgative filter**

---

[16]. See Daniel C. Dennett, *Darwin's Dangerous Idea: Evolution and the Meanings of Life* (New York: Simon & Schuster, 2014) for the conceptual framework linking cosmic dynamics to evolutionary processes. Dennett's "universal acid" of the algorithmic process is what NPN identifies as the operation of the *Logos*.

CHAPTER 4. NPN METAPHYSICS 47

of natural selection. This is the force that systematically removes maladaptive "dead ends." Speciation is the creative act of **Neikos** (Νεῖκος); extinction is its purgative one.

An organism itself is a temporary, localized equilibrium of these forces. It is a sustained act of **Neikos** (Νεῖκος) (maintaining a bounded, individuated structure against entropy) that is utterly dependent on processes of **Philia** (Φιλία) (the chemical bonds of its molecules, the symbiotic cooperation of its cells). This dynamic is not confined to biology; it is observable in the formation of stars, the dynamics of ecosystems, and the rise and fall of societies. It is the **universal dialectic of the *Archē*** (Ἀρχή).

The "assessment of Time" is this entire, impartial, diachronic **Becoming**. It is the metric that judges systems not by their momentary existence, but by their sustained capacity to navigate the eternal tension between cohesive inheritance and selective pressure. All change,stability, and complexity are manifestations of **Physis** (Φύσις)—the dynamic balance of this eternal tension.

4.6.7 The Evidence in Classification: A Snapshot of the Dance

The power of this framework is that it makes sense of not only processes but also their products. The scientific act of **biological classification (taxonomy)** provides a stunning snapshot of this eternal dance.

When a biologist groups organisms into a species or genus, they are mapping the historical work of **Philia**. The shared, inherited traits—the common skeletal structure, the conserved genetic sequence—are the physical signature of *Philia*'s cohesive force, preserving a shared "blueprint" across deep time.[17]

Conversely, the traits that are absent, the diagnostic differences that separate one group from another, are the recorded work of **Neikos**. The mutations that were *not* shared, the adaptations that made one lineage distinct,

---

17. This cohesive, historical dimension of biological kinds is central to the "phylogenetic species concept," which defines taxa by shared ancestry. See Ernst Mayr, *Systematics and the Origin of Species* (New York: Columbia University Press, 1942), 120–21.

are the literal marks of *Neikos* as the force of separation and individuation.[18] Furthermore, the vast space of possible traits that a lineage *could* have—but does not—bears the silent, negative imprint of *Neikos*. Every maladaptive or less optimal variation that was selected against, every potential form that failed the test of time, represents the purgative, filtering action of this same force, systematically removing what does not work from the historical record.

A taxonomic tree is thus a frozen record of the interplay between *Philia* (which builds the branches through shared ancestry) and *Neikos* (which splits them through speciation and prunes them through extinction).

4.7 Conclusion: The Laws of the Territory

This foundational section completes the first and most critical component of the Neo-Pre-Platonic Naturalist meta-structure: the definition of the **objective territory** to be navigated.

The **Diachronic Turn** has established that the **First Principle of Becoming** is ontologically primary, revealing the *Archē* not as a static substance, but as a lawful, dynamic process—***Hylē*-as-actualized-by-*Logos***. Its entire generative capacity is driven by the exhaustive polarity of two fundamental, amoral cosmic forces: *Philia* (Union) and *Neikos* (Separation). This constant, impartial dynamic is the **universal dialectic** that structures all emergent complexity, from the orbit of a star to the genetic code of an organism—with the entirety of the *Archē* bounded by the silent, boundless *Apeiron*.

We have now established the laws of the universe. But the ultimate challenge of NPN is not to merely describe the territory; it is to provide a coherent account of the **Navigator** that lives and strives within it. If the cosmic forces of *Philia* and *Neikos* drive all existence, they must also constitute the internal architecture of the mind.

---

18. Darwin's "principle of divergence" describes how natural selection drives lineages apart, a process of differentiation that mirrors the *Neikos* dynamic. See Charles Darwin, *On the Origin of Species* (London: John Murray, 1859), 111–26.

CHAPTER 4. NPN METAPHYSICS

The exploration of the objective *Archē* is complete.[19] The stage is now set for the next two chapters: to systematically derive the **First Principles of the Emergent Agent**—the stratified *Psyche* and its foundational engine of striving, the *Hormē*.[20]

---

19. This conception of a lawful, dynamic *Archē* in which striving (*Hormē*) is constitutive resonates with Spinoza's metaphysics of *conatus*—the inherent striving of each thing to persist in its being. NPN elevates *conatus* from a psychological observation to a cosmological necessity. See Benedict de Spinoza, *Ethics*, in *The Collected Works of Spinoza*, vol. 1, trans. Edwin Curley (Princeton, NJ: Princeton University Press, 1994), Part III, Prop. 6–7.

20. The view of existence as fundamentally creative, dynamic, and generative—where novelty emerges through time—is central to Henri Bergson's *Creative Evolution*. NPN adopts Bergson's "Duration" but rejects *Élan Vital* in favor of the thermodynamic *Hormē*. See Henri Bergson, *Creative Evolution*, trans. Arthur Mitchell (New York: Henry Holt and Company, 1911), 1–10, 248–73.

Chapter 5

The Engine of Agency: The Constitutive *Hormē* and the Origin of Value

"You could not discover the limits of the soul, even if you traveled every road; so deep is its Logos." — Heraclitus[1]

5.1 Introduction:
From Cosmic Laws to the Conscious Navigator

The previous chapter established the metaphysical bedrock of Neo-Pre-Platonic Naturalism. We have a complete map of the territory: a dynamic *Archē* (Ἀρχή) whose fundamental substance is *Hylē*-as-actualized-by-*Logos*, a process driven by the exhaustive polarity of *Philia* and *Neikos* and bounded by the silent *Apeiron*. This is a powerful description of an objective, lawful, and creative cosmos.

But a map is inert. A territory, no matter how well-charted, requires a map-reader. This presents the next, more intimate crisis for our first-principles framework: if reality is an impersonal, causal process, **why does anything care what happens next?**

The philosophical tradition has often struggled to explain how value, purpose, and striving can arise from a universe of indifferent atoms. This has created an unbridgeable chasm between the world of facts (the domain of science) and the world of value (the domain of ethics).

---

[1]. Establishes the foundational premise that the *Logos* is a pervasive, deep structure within the psyche, not merely a surface phenomenon. Heraclitus, Fragment 45 (KRS 235), in G. S. Kirk, J. E. Raven, and M. Schofield, *The Presocratic Philosophers: A Critical History with a Selection of Texts*, 2nd ed. (Cambridge: Cambridge University Press, 1983), 195.

This chapter bridges that chasm not by adding a supernatural soul, but by identifying the **thermodynamic engine** inherent in life itself. We will demonstrate that the transition from object to agent is not a mystical leap, but a physical one. It is the emergence of the *Hormē* (Ὁρμή)—the constitutive, non-negotiable impulse to persist against the flow of entropy.

In this chapter, we will ground agency directly in the physics of the *Archē*. We will solve the "Is-Ought" problem by showing that for a striving system, the "Ought" is simply the functional requirement of its continued existence. Before we can analyze the *structure* of the mind, we must first understand the *engine* that powers it.

5.2 The Constitutive *Hormē* (Ὁρμή): Agency as Striving

"Nature does nothing in vain." — Aristotle[2]

"Life is that which can mistake." — Jacob von Uexküll[3]

The transition from the inert *Archē* to the conscious Navigator requires a bridge. We have defined the laws of the territory, but a territory does not navigate itself. What force animates the system? What impulse drives the *Orexis* (Ὄρεξις), motivates the *Thymos* (Θυμός), and demands the *Logistikon* (Λογιστικόν)? This force is not a secondary component but the constitutive ground of the entire living system: the *Hormē* (Ὁρμή)—the innate, non-negotiable impulse to strive, to persist, and to flourish.[4]

In this section, we ground the concept of agency not in metaphysics, but in the observable physics of the *Archē*. We demonstrate that agency is a

---

2. Classical grounding for the teleological functionality observed in natural systems. Aristotle, *De Caelo (On the Heavens)*, Book I, 271a33, in *The Complete Works of Aristotle: The Revised Oxford Translation*, ed. Jonathan Barnes, vol. 1 (Princeton: Princeton University Press, 1984), 448.

3. Defines agency by the capacity for error, implying an internal standard of success/-failure (the *Hormē*) absent in inert matter. Cited in Giorgio Agamben, *The Open: Man and Animal*, trans. Kevin Attell (Stanford: Stanford University Press, 2004), 39–40.

4. Establishes the biological basis of self-maintenance (autopoiesis) as the root of cognition. Humberto R. Maturana and Francisco J. Varela, *Autopoiesis and Cognition: The Realization of the Living* (Dordrecht: D. Reidel Publishing Co., 1980), 73–123.

CHAPTER 5. AGENCY: HORMĒ AND THE ORIGIN OF VALUE

specific causal orientation and that objective value is its necessary functional output.

5.2.1 The Thermodynamic Basis of Agency

Agency is not a mysterious addition to matter; it is the defining characteristic of a system that actively resists entropy. To be an agent is to be a system organized to maintain its own existence against the dissolving forces of *Neikos* (Νεῖκος).

Consider the simplest bacterium swimming up a chemical gradient toward nutrients. It is not merely reacting; it is expressing the primordial *Hormē*. Its molecular machinery encodes a primitive valuation: "this direction supports persistence; that direction threatens it."[5] This is not consciousness, but it is indisputably agency—the capacity to bias outcomes toward survival.

> **First Principle**
>
> **FP6: FIRST PRINCIPLE OF AGENCY:**
> **THE PRIMACY OF THE HORMĒ**
>
> **Statement:** The *Hormē* is the constitutive, non-negotiable ground of being an agent.
>
> **Justification:** The negation is a functional absurdity that violates the causal structure of reality. To assert that an agent exists without *Hormē* is to claim that an inert system can persist against the entropic forces of *Neikos*. In a dynamic *Archē*, any system that fails to strive is strictly identical to a decaying object. Therefore, it is a performative contradiction to assert that an agent—an entity defined by its persistence—can exist without the striving necessary to endure through deep time.

---

5. Empirical evidence of biased movement (agency) at the unicellular level. Howard C. Berg, *Random Walks in Biology*, expanded ed. (Princeton: Princeton University Press, 1993), 92–112.

A system without intrinsic *Hormē* is, by strict definition, an object. It may be moved, but it does not move *itself* in service of an endogenous aim. The *Hormē* is the active principle that separates the living from the dead; it is the thermodynamic work of maintaining a far-from-equilibrium state.[6]

5.2.2 The Objective Ground of Value

From the constitutive reality of the *Hormē*, the objectivity of value follows with logical necessity. Value is not a subjective projection, nor is it a binary preference for one cosmic force over another. **Philia** (Φιλία) and **Neikos** are amoral drivers; neither is inherently "good" or "bad".

- 'Good' is the state of **Dikaiosynē** (Δικαιοσύνη)—the functional alignment and dynamic balance of these forces that fulfills the system's striving within the *Logos*.
- 'Bad' is the state of misalignment or imbalance that frustrates that striving.

For a complex organism, social cohesion (*Philia*) is not a static good, nor is individuation (*Neikos*) a static bad. Excessive cohesion leads to stagnation and the death of the agent's distinct identity; excessive separation leads to fragmentation and the loss of supportive structures. **Value is discovered** in the timely, proportional balancing of these forces to achieve optimal alignment with reality.[7]

5.2.3 Agency as the Redirection of Causal Flow

In a lawful *Archē* (FP2), what distinguishes an agent from a deterministic object? The difference lies not in escaping causality, but in the *nature of the causal pathway*. **An agent is a subsystem organized to redirect causal flows toward outcomes that support its own persistence.**

---

6. Thermodynamic justification for why life requires constant work (striving) to resist entropy. Jeremy L. England, "Dissipative Adaptation in Driven Self-Assembly," *Nature Nanotechnology* 10, no. 11 (2015): 919–23. For the complete derivation of Theorem T7 (The Entropic Asymmetry), see Appendix B.

7. Bridges biological function with moral realism; value is a factual claim about what sustains a life form. Ruth Garrett Millikan, *Language, Thought, and Other Biological Categories* (Cambridge, MA: MIT Press, 1984), 1–30; Peter Railton, "Moral Realism," *The Philosophical Review* 95, no. 2 (1986): 163–207.

When an animal avoids a trap, it alters the behavioral sequence that would have led to its death. The system's internal organization—shaped by and for the *Hormē*—introduces a *teleological bias* into the causal matrix.[8] The agent acts as a local source of work against entropic dispersion, carving a trajectory of continued existence out of a field of possible decay.

This capacity scales with complexity. The human *Nous* (Νοῦς), with its ability to model the future, can redirect causal flows across decades. But the root logic remains identical to the bacterium: action in service of the *Hormē*, guided by an alignment—whether instinctual or calculated—with the *Logos*.

5.2.4 Genetic Memory: The Archē's Hard-Won Logos

The *Hormē* does not operate in a vacuum. Across deep time, successful strategies for persistence are winnowed by *Neikos* (selection) and preserved by *Philia* (inheritance). An organism's genetic code is a **compressed historical record** of these successful strategies.[9]

This "genetic knowledge" manifests as instinct, adaptation, and behavioral predisposition. The infant's grasp, the fear of sudden shadows, the drive to bond—these are expressions of the ancestral *Hormē* crystallized into biological form. They are the pre-conscious precursors to what we will identify in the next chapter as the **Somatic logos** (Σωματικός λόγος)—the logic of the body.

With the engine of the *Hormē* defined, we must now examine the specific machinery human beings have evolved to serve it. We turn now to the architecture of the Navigator: the Stratified *Psyche*.

5.2.5 The Derivation of Objective Value

The stratified model of the *Psyche*, grounded in the constitutive *Hormē*, provides the foundation for a decisive resolution to the problem of value.

---

8. Explains how teleology (purpose) can exist within a physicalist system via constraints on work. Deacon, *Incomplete Nature*, 267–310.

9. Conceptualizes the genome as a data-store of past survival strategies. Dawkins, *The Selfish Gene*, 1–21.

Objective value is not an invented concept but a deduced feature of any striving system within a lawful *Archē*.

The Deduction

1. **The First Principle of the *Hormē*:** Any system for which the concept of 'value' is meaningful must be a system that strives. The *Hormē* is the constitutive, non-negotiable ground of being an agent.
2. **The Functional Definition of Value:** Given a system with *Hormē*, 'good' is the **functional state of alignment** between the system's condition/actions and the fulfillment of its striving. 'Bad' is the state of misalignment and frustration. For a living organism, nourishment is *objectively good* and poison is *objectively bad* relative to its *Hormē*. This is a functional fact, not an opinion.
3. **The Role of the *Logos*:** The *Archē* is a lawful framework. Therefore, what *actually* fulfills or frustrates the *Hormē* is determined by the causal structure of reality. The *Logos* provides the objective, impersonal standard against which strategies for fulfilling the *Hormē* are judged by the cosmic ***Dikē*** (Δίκη)—the impartial justice of the *Archē* itself.

Implications: From Description to Prescription

This derivation transforms value from a matter of taste into a feature of natural philosophy.

- **Value Claims are Functional Claims:** To make a value claim is to make a claim about which states and actions most effectively fulfill a system's *Hormē* within the *Logos*.
- **A Navigational Tool:** This does not provide a list of rules but a **methodology**. It grounds the Socratic project, turning the search for the good life into the practical task of aligning one's Somatic *logos* with the external *Logos*.

In Neo-Pre-Platonic Naturalism (NPN), the **Humeian Gap** (the "Is-Ought" problem) is identified as a synchronic illusion generated by analyzing reality in frozen, timeless snapshots. By executing a **Diachronic**

CHAPTER 5. AGENCY: HORMĒ AND THE ORIGIN OF VALUE

Turn, NPN demonstrates that for any system that strives to persist, the "Is" and the "Ought" are functionally identical.

> **Corollary**
>
> **C4: PRIMARY COROLLARY: THE OBJECTIVITY OF VALUE**
>
> **Statement:** For any system possessing a constitutive *Hormē*, 'good' is the functional state of alignment (Dikaiosynē) that fulfills its striving, and 'bad' is the state of misalignment that frustrates it. Value is an objective relationship between a system's dynamic balance and its successful expression within the constraints of the *Logos*.
>
> **Justification:** This is the direct ethical consequence of **FP6**. To deny this is to assert there is no functional difference between a flourishing and a dying system, which is empirically false and logically absurd.

5.2.6 The Dissolution of the Humeian Gap

> *"In every system of morality, which I have hitherto met with, I have always remarked, that the author proceeds for some time in the ordinary way of reasoning... when of a sudden I am surprised to find, that instead of the usual copulations of propositions, **is**, and **is not**, I meet with no proposition that is not connected with an **ought**, or an **ought not**."* — David Hume[10]

The traditional gap between descriptive facts ("Is") and prescriptive values ("Ought") collapses when agency is grounded in the **Hormē (Ορμή)**—the innate impulse to strive and persist.

---

10. David Hume, *A Treatise of Human Nature* (1739), 3.1.1. This passage establishes the "Humeian Guillotine" that severed fact from value. NPN asserts that this gap only exists in a synchronic analysis; once the diachronic engine of *Hormē* is introduced, the "Is" of the living agent constitutively necessitates the "Ought" of its own survival.

- **The Is-Ought Unity**: For a living system, the "Ought" is the operational output of its "Is". To *be* an agent that persists across time is to be a system governed by the imperative to act in alignment with the **Logos (Λόγος)**.
- **The Synchronic Illusion**: Synchronic analysis views "Is" as a static fact devoid of value. In a diachronic, process-primary reality, however, an "Is" is an ongoing act of **Becoming** that requires constant work to maintain against entropic forces.
- **Hormē as the Bridge**: The Hormē is the transcendental condition for valuation. Because a system must strive to navigate the **Archē (Ἀρχή)** successfully or face extinction, the question "Why ought I to strive?" is biologically and logically incoherent—it is the very meaning of being an agent.

Value is not a subjective project but a functional relationship discovered between a striving navigator and its environment.

- **Objective "Good"**: This is defined as the functional state of alignment (Dikaiosynē) that fulfills the system's striving.
- **Objective "Bad"**: This is the state of misalignment or imbalance that frustrates the Hormē.
- **Dikē (Δίκη)**: The Logos provides the objective standard against which these strategies are judged by the impartial, causal justice of reality.

> "We must know that war is common to all and strife is justice, and that all things come into being and pass away through strife."
>
> —Heraclitus[11]

---

11. Heraclitus, Fragment DK 22 B80. This fragment establishes the ontological necessity of *Neikos*. Justice (*Dikē*) is not static peace, but dynamic tension. This anticipates **Theorem T5**, confirming that to suppress opposition is to violate the fundamental structure of the *Archē*.

> **Theorem**
>
> ### T5: THEOREM: THE ENTROPIC MANDATE
>
> **Statement:** Any system that attempts to permanently suppress *Neikos* (dissent, variation, friction) in favor of pure *Philia* (unity, stability) guarantees its own entropic collapse. Stability is not a static state, but a dynamic oscillation.
>
> **Derivation:** This theorem applies cosmic dynamics to political and systems theory.
>
> 1. A system strives to persist (**FP6: Primacy of the Hormē**), but the environment (*Archē*) is constantly changing (**FP2: Diachronic Primacy**).
> 2. Persistence therefore requires adaptation.
> 3. Adaptation requires *Neikos* (differentiation/selection) to break down maladaptive structures and generate new options (**FP3: Logos and Polarity**).
> 4. A system that eliminates *Neikos* eliminates its ability to adapt to the changing *Archē*.
>
> **Conclusion:** "Perfect order" (Pure *Philia*) is functionally identical to death. A healthy, persisting system *must* institutionalize controlled *Neikos*.

5.3 Ethical Isomorphism

Because "Truth" is the alignment of internal models with the Logos and "Good" is the alignment of actions with that same Logos, the pursuit of Truth and the pursuit of the Good are the same mechanical operation (Theorem T4). An "Ought" is simply the requirement for a Navigator to act upon a map that matches the territory of the Archē.

> **Theorem**
>
> ### T4: THEOREM: ETHICAL ISOMORPHISM
>
> **Statement:** Epistemic error (falsehood) and Ethical vice (immorality) are functionally isomorphic; both are states of misalignment between the *Navigator's* internal models and the external *Logos*. Therefore, the pursuit of Truth and the pursuit of the Good are the same mechanical operation.
>
> **Derivation:** This theorem unifies the epistemic and ethical branches of the system.
>
> 1. **Definition of Good:** "Good" is defined as the functional alignment of a system's actions with the *Logos* to fulfill its *Hormē* (**C4: The Objectivity of Value**).
> 2. **Definition of Truth:** "Truth" (or valid reasoning) is the functional alignment of the *Nous's* logic with the causal structure of the *Archē* (**FP7: The Somatic Logos**).
> 3. Since both Good and Truth are defined by the single metric of *Functional Alignment with the Logos*, a failure in one is necessarily a failure in the other.
> 4. Maladaptive behavior (vice) is the result of acting on a map that does not match the territory.
>
> **Conclusion:** An unethical act is an act based on a false model of reality. To improve character is to improve one's map.

5.4 The Life-Agency Isomorphism

Ultimately, we must dissolve the artificial distinction between biological existence and navigational agency. The ability to navigate—to perceive a difference and act to minimize friction—is not a higher-order faculty added to life; it is the definition of life itself.

This leads us to the **Life-Agency Isomorphism**: A system is an agent if and only if it possesses *Hormē* (the thermodynamic striving to persist). Therefore, agency is not a psychological phenomenon, but a physical one. To be alive is to perform work against entropy; to perform that work is to be an agent.

This establishes a critical ontological boundary between **Agents** and **Artifacts**. A machine or a tool may possess sophisticated functions, but it lacks *Hormē*. Its persistence is not causally linked to its performance; if a calculator calculates incorrectly, it does not "die." It has a function, but it does not have a stake in the outcome. The Navigator, by contrast, is a system where performance and persistence are coupled. Failure to navigate the terrain correctly results in the thermodynamic dissolution of the agent itself.

Consequently, this isomorphism provides the rigorous grounding for our ethics and resolves the "Is-Ought" problem. In a universe of dead matter, "facts" and "values" are indeed separate categories. But in a living system, the "Fact" of existence is inherently directional. To *be* (as a far-from-equilibrium system) is to *ought to persist*.

Value, therefore, is not a mystical layer imposed on the universe; it is generated by the metabolic requirements of the agent. The "Ought" is simply the set of actions required to sustain the "Is." From the bacterial struggle up a sugar gradient to the complex moral deliberations of the human *Nous*, the logic remains scale-invariant: the Good is that which aligns with the continuation of the pattern. See Appendix B for the formal derivation of Theorem T6.

5.5 Conclusion: The Imperative to Navigate

We have established the bedrock of the Navigator's existence. The *Archē* is not a static stage, but a dynamic arena. Within this arena, the *Hormē* is not a choice, but a constitutive necessity—the thermodynamic engine that separates the agent from the entropic drift of the object.

From this engine, we have derived the most critical ethical breakthrough of the NPN system: **The Objectivity of Value (C4)**. Good and Bad are not matters of opinion; they are functional descriptions of alignment and misalignment. To flourish (*Eudaimonia*) is to successfully navigate the causal currents of the *Logos* in service of the *Hormē*.

But a raw engine cannot steer. A blind drive, no matter how powerful, will inevitably crash against the hard constraints of reality. To convert this raw *Hormē* into sustained *Energeia*, the organism requires more than just impulse; it requires instruments. It needs a sensor array to detect the territory (*Aisthesis*), a computer to calculate trajectories (*Logistikon*), and a command center to decide the course (*Nous*).

We must now turn our gaze from the engine to the vehicle. How did the blind striving of the *Hormē* evolve into the lucid insight of the Navigator? The answer lies in the structural evolution of the *Psyche*.

Table 5.1   The Structure of the *Psyche*: Lexicon for Chapter 6

NPN Term	Role	Technical Definition
Orexis (ὄρεξις)	Instinct	The domain of immediate instinct; the most ancient, short-term layer of the *Hormē* (e.g., hunger, fear).
Thymos (θυμός)	Strategy	The strategic social calculator; the mid-term layer of the *Psyche* managing social conflict and cohesion.
Logistikon (λογιστικόν)	Executive	The calculating faculty of the *Nous* that orchestrates *Orexis* and *Thymos* to achieve long-term goals.

Chapter 6

The Architecture of the Navigator: From Impulse to Insight

"The mind is a system of organs of computation, designed by natural selection to solve the kinds of problems our ancestors faced." — Steven Pinker[1]

6.1 Introduction: The Evolutionary Layering of the *Hormē*

We have established that the *Hormē* is the engine of agency—the raw drive to persist. But a raw engine cannot steer. A blind drive, no matter how powerful, will inevitably crash against the hard constraints of the *Logos*. To convert raw striving into sustained *Energeia*, the organism requires instruments.

The philosophical tradition has often treated the knowing subject—the mind, the self, consciousness—as a spectral exception to the natural order. This chapter executes the decisive inversion required to resolve this dualism: **the knower is not a mysterious exception to the *Archē*, but its most sophisticated known product.**

---

[1]. Steven Pinker, *How the Mind Works* (New York: W. W. Norton & Company, 1997), 21–30. NPN accepts the computational model but grounds it in the thermodynamic imperative of the *Hormē*.

We will demonstrate that the human *Psyche* (Ψυχή)—with its layered architecture and capacity for reason—is a perfectly natural, **emergent phenomenon**.² Its structure is a palimpsest of evolutionary strategies, a layered record of how the *Hormē* has adapted to navigate the *Archē* over different time scales, from the immediate reflexes of the *Orexis* to the multidecade and even intergenerational planning of the *Nous*.

While the *Hormē* provides the singular engine, it cannot navigate a complex environment as a monolith. To persist against the **Entropic Mandate (T5)**, it must stratify.³

In this chapter, we will map this architecture—from the immediate biochemical urgency of **Orexis**, to the social calculations of **Thymos**, and finally to the emergent executive power of the **Nous**. We will discover that the Navigator is not a "ghost in the machine," but the machine's own emergent capacity to steer itself.

6.2 The Stratified *Psyche*: An Evolutionary Layering of the *Hormē*

The complex human mind did not emerge *de novo*. Its structure is a palimpsest of evolutionary strategies, a layered record of how the *Hormē* has adapted to the increasing complexity of the *Archē*. While the *Hormē* provides the singular engine of agency, it cannot navigate a complex environment as a monolith. To persist against the Entropic Mandate (T5), it must stratify.⁴

Stratification is the evolutionary response to the difficulty of maintaining the balance between *Philia* (cohesion) and *Neikos* (differentiation). As the

---

2. Provides the physicalist framework for how higher-order teleology arises from thermodynamic processes, a concept NPN adapts as "Ententionality." Terrence W. Deacon, *Incomplete Nature: How Mind Emerged from Matter* (New York: W. W. Norton & Company, 2012), 1–37, 267–310.

3. A precursor to the concept of the *Nous* organizing conflicting drives. NPN identifies this as the expression of the *Will to Power* (*Hormē*) stratifying to manage complexity. Friedrich Nietzsche, *The Will to Power*, ed. Walter Kaufmann (New York: Vintage, 1968).

4. A precursor to the concept of the *Nous* organizing conflicting drives. NPN identifies this as the expression of the *Will to Power* (Hormē) stratifying to manage complexity. Friedrich Nietzsche, *The Will to Power*, ed. Walter Kaufmann (New York: Vintage, 1968).

organism faces more dynamic threats and opportunities, the *Hormē* differentiates into specialized subsystems, each designed to manage a specific horizon of time and complexity.

6.2.1 The Quadripartite Architecture

The human *Psyche* is the result of this evolutionary layering—a functional hierarchy where lower layers manage immediate necessity and higher layers manage long-term probability.

- **Orexis (Ὄρεξις): The Domain of Instinct.** The most ancient, biochemical layer. It manages immediate homeostatic needs such as hunger, thirst, pain, and libido. It operates on the shortest time horizon: the immediate now.
- **Thymos (Θυμός): The Social Calculator.** The layer of strategic emotion. It manages in-group/out-group dynamics, status, and social cohesion. It translates the raw drive of *Hormē* into social vectors like pride, shame, and loyalty.
- **Logistikon (Λογιστικόν): The Executive Processor.** The interface that mediates between the demands of the lower drives and the reality of the external world. It orchestrates the *Orexis* and *Thymos* to achieve coherent action, preventing internal fragmentation.
- **Nous: The Emergent Navigator.** The apex faculty capable of abstract model-building, self-reflection, and conscious decision-making (*Prohairesis* [Προαίρεσις]). It is the only faculty capable of modeling the *Archē* across deep time.

6.2.2 The Functionalist Defense of Consciousness

A system lacking this stratified architecture would be unable to navigate the *Archē* and would inevitably succumb to extinction. Therefore, consciousness is not defined by a "ghostly" substance, but by its function: the *Energeia* (Ἐνέργεια) of a system successfully modeling its own relation to the *Logos*.

This functional alignment is the answer to the "Hard Problem" of meaning. In the NPN framework, semantics (meaning) emerges when syntax (the

Logistikon) is driven by *Hormē* (striving). A system that *must* navigate to survive does not merely process symbols; it processes stakes.[5]

6.2.3 Orexis: The Domain of Immediate Instinct

This is the most ancient layer, the direct biochemical expression of the *Hormē* as it pertains to immediate survival.[6]

- **Neikos-driven Orexis**: The impulse to separate from harm—fleeing a predator, recoiling from pain.
- **Philia-driven Orexis**: The impulse to unite with a resource—the drive to eat, to drink, to mate.
- **Function**: *Orexis* is the engine of homeostatic regulation, a suite of hardcoded and conditioned reflexes for managing the body's immediate state within the *Archē*.

6.2.4 Thymos: The Strategic Social Calculator

As organisms became social, a more nuanced system was required to navigate the complex in-group/out-group dynamics.

- **Neikos-driven Thymos**: Anger at a rival, courage against a threat, jealousy—drives related to strategic separation and competition.
- **Philia-driven Thymos**: Pride in one's group, shame, empathy, loyalty—drives related to strategic attraction, bonding, and maintaining social cohesion.
- **Function**: *Thymos* manages the mid-term strategic landscape of status, belonging, and coalition-building, all in service of the individual's and group's *Hormē*.

---

5. Defends the position that syntax alone (computation) is insufficient for semantics (meaning). NPN argues that *Hormē* (the will to survive) is the missing ingredient that grounds syntax in reality, resolving the "Chinese Room" paradox. John Searle, "Minds, Brains, and Programs," *Behavioral and Brain Sciences* 3, no. 3 (1980): 417–57.

6. Supports the view that cognition is grounded in sensorimotor systems rather than abstract symbols. Lawrence W. Barsalou, "Perceptual Symbol Systems," *Behavioral and Brain Sciences* 22, no. 4 (1999): 577–609.

6.2.5 The Necessary Emergence of the *Logistikon*

The stratification of the *Psyche* is not arbitrary. The emergence of the *Thymos* created a new problem for the system: it generated a flood of complex, often contradictory, social data (alliances, threats, status). A simple reactive drive (*Orexis*) was no longer sufficient. To resolve these conflicts and produce *coherent social action*—action that could successfully fulfill the *Hormē* within a social group—a new subsystem was evolutionarily mandated.

This subsystem is the **Logistikon**. It is not a lucky accident but a **functional necessity**. Without a central processor to integrate the signals of *Orexis* and *Thymos*, an organism would be trapped in a chaotic loop of un-orchestrated impulses, fundamentally failing its core *Hormē*. The *Logistikon* is the necessary navigator for the social territory that the *Thymos* reveals.[7] From this primary function of coordination, its higher capacities for abstract model-building and long-term strategy later emerged.

6.2.6 The Emergent *Nous* and the *Logistikon*

Only with this foundational machinery in place could the capacity for abstract modeling—the *Nous*—emerge. Its executive function, the *Logistikon*, does not replace the lower layers but orchestrates them.

- **Function of the Logistikon**: To model the long-term *Logos* of the environment and the self, and to strategically deploy or suppress *Orexis* and *Thymos* to achieve goals that fulfill the *Hormē*[8] across extended time horizons.[9]

---

7. The dual-system theory aligns with the distinction between the automated *Orexis/Thymos* and the deliberative *Logistikon*. Daniel Kahneman, *Thinking, Fast and Slow* (New York: Farrar, Straus and Giroux, 2011), 105–18.

8. Cognitive science support for the *Somatic logos*, showing how abstract reason is metaphorically grounded in bodily experience. Barsalou, "Perceptual Symbol Systems," 577–95; George Lakoff and Mark Johnson, *Philosophy in the Flesh: The Embodied Mind and Its Challenge to Western Thought* (New York: Basic Books, 1999), 3–20, 77–105.

9. The original model of the tripartite soul from which NPN derives the functional stratification. NPN naturalizes this architecture, stripping it of its metaphysical dualism. Plato, *Republic*, Book IV, 439d–441c.

- **The State of Dikaiosynē**: This is the functional harmony of the entire system, where the *Logistikon* successfully governs the *Psyche*, aligning the immediate and social expressions of the *Hormē* with long-term, rational ends that are in alignment with the Logos.

Table 6.1 The Stratified Psyche: Evolutionary Layers of Agency

Layer	Modern Correlate	Primary Drive	Function
Orexis (Instinct)	Autonomic	Survival (Bio-Chemical)	Immediate homeostatic regulation (e.g., hunger, pain response).
Thymos (Spirit)	Social Intuition	Status & Belonging	Strategic social calculator; manages in-group/out-group dynamics.
Logistikon (Executive)	Executive Function	Coherence & Planning	The interface that orchestrates lower drives to achieve long-term goals.
Nous (Navigator)	Conscious Awareness / Meta-Cognition	Meaning & Alignment	The emergent integrator; capable of modeling the self and the *Archē*.

"Let us liken the soul to the composite nature of a pair of winged horses and a charioteer... One of them is noble and good, and of good stock, while the other is the opposite and of opposite stock. And so the driving in our case is necessarily difficult and troublesome." — Plato[10]

---

10. Plato, Phaedrus, 246a–b. In the NPN framework, the Charioteer represents the Logistikon (Executive Function) attempting to navigate the Archē. The noble horse represents Philia-driven Thymos (honor, loyalty, cohesion), while the unruly horse represents Neikos-driven Orexis (appetite, impulse, self-interest). The "difficulty" Plato describes is the thermodynamic work required to maintain Dikaiosynē (functional harmony) against the entropic tendency of the drives to separate.

CHAPTER 6. ARCHITECTURE OF THE NAVIGATOR

> **First Principle**
>
> **FP4: FIRST PRINCIPLE OF EMERGENT COMPLEXITY: THE POTENTIAL FOR *NOUS***
>
> **Statement:** The constitutive interaction of *Philia* and *Neikos* within the *Logos* is inherently generative, making the emergence of complex, stratified systems—including the *Nous*—a natural and potential outcome of the *Archē*'s dynamics, not a supernatural accident.
>
> **Justification:** The negation leads to a double contradiction.
>
> 1. First, an **ontological contradiction**: Since the *Archē* is the sole fundamental reality (FP1), any phenomenon that exists must have its sufficient cause within the *Archē*. The *Nous* exists; therefore, the generative potential for the *Nous* must be a constitutive property of the *Archē*.
>
> 2. Second, a **performative contradiction**: To claim that the *Archē* cannot produce the *Nous* is to use the *Nous* to deny its own origin. It is the act of a product denying the capacity of its own factory. To argue the point is to prove it false.

6.2.7 The Reflexive Navigator: Meta-Cognition and Self-Modeling

The final layering of the *Psyche* (Ψυχή) introduces a capacity unique to the **Nous** (Νοῦς): the ability to turn the modeling engine inward. While the *Logistikon* (Λογιστικόν) models the external *Logos* (Λόγος) to calculate trajectories, the *Nous* performs a reflexive operation, modeling the Navigator itself. This **Meta-Cognition** is not a luxury of abstract thought but a navigational requirement.

For a system to self-correct, it must be able to represent its own internal states and models as objects of inquiry. Without this meta-cognitive potential, the system would be trapped in its current programming, unable to recognize the failure of a model or the presence of unexamined ***Doxa*** (Δόξα). The *Nous* provides the "view from above" that allows the Navigator to evaluate the accuracy of its own compass.

> **First Principle**
>
> ### FP9: FIRST PRINCIPLE OF THE NOUS: META-COGNITIVE POTENTIAL
>
> **Statement:** The *Nous* is inherently capable of reflexive self-modeling. It can form models not only of the external *Archē* but also of its own processes, states, and models. This self-referential capacity is not merely an observed property but a *transcendental condition* for any system that can formulate propositions about its own nature.
>
> **Justification:** The negation is a performative contradiction. To claim "the *Nous* cannot model itself" is to *present a model of the Nous*. The act of denial is itself an instance of the capacity being denied.

6.3 The Topography of Navigation: Mapping Consciousness and Reaction

The stratified *Psyche* operates not only through different drives but through distinct modes of awareness. The NPN framework maps the modern distinction between **conscious volition** and **unconscious reaction** onto the ancient tripartite structure, revealing the *Nous* not as a detached spirit, but as the emergent **Integrator** of a complex, automated system.

6.3.1 The Substrate: *Orexis* as Biological Instinct

- **Mode:** Purely Reactive / Hardwired.
- **Modern Correlate:** The Autonomic and Limbic baseline (System 1)[11]

---

11. Identifies the biological substrate of the ancient 'appetite'. Kahneman, *Thinking, Fast and Slow*.

CHAPTER 6. ARCHITECTURE OF THE NAVIGATOR 71

- **Function:** *Orexis* operates entirely below the threshold of the *navigator's* conscious awareness. It is the domain of **Instinct**. It monitors the biochemical *Hylē* ('Υλη) and generates immediate, non-negotiable impulses (hunger, pain, lust). It is the raw *Hormē* ensuring metabolic persistence.

6.3.2 The Filter: *Thymos* as the Social Unconscious

- **Mode:** Automated Heuristic / "Hot" Cognition.
- **Modern Correlate:** The Social Unconscious / Intuition (System 1)[12]
- **Function:** *Thymos* is the repository of internalized social conditioning and deep-seated emotional strategies. While we feel its output (anger, shame, pride), its *processing* is unconscious. It is a sophisticated, high-speed calculator that scans the social environment for threats (*Neikos*) and alliances (*Philia*) and presents the *Nous* with a pre-packaged emotional conclusion.

6.3.3 The Interface: *Logistikon* as the Orchestrator

- **Mode:** Analytical / Algorithmic.
- **Modern Correlate:** Executive Function / Simulation Engine (System 2)[13]
- **Function:** The *Logistikon* is not the "self," but the **instrument** of the self. It acts as the **Orchestrator** between the raw data of the lower faculties and the awareness of the *Nous*. It is the interface that translates the "hot" signals of *Orexis* and *Thymos* and the raw data of *Aisthēsis* (Αἴσθησις) into "cool," coherent models.[14] It acts as

---

12. Demonstrates that emotion (Thymos) is critical for rational decision-making, not an impediment. Antonio Damasio, *Descartes' Error: Emotion, Reason, and the Human Brain* (New York: G. P. Putnam's Sons, 1994).

13. Maps the *Logistikon* to the effortful, calculating mode of thought necessary for complex problem solving. Kahneman, *Thinking, Fast and Slow*, 105–18.

14. This translation of raw data into coherent models mirrors Immanuel Kant's distinction between *Phenomena* (things as structured by the mind) and *Noumena* (things-in-themselves). NPN reinterprets Kant's "Transcendental Idealism" as a form of "Transcendental Realism with a User Interface": the *Noumena* exists (FP1), but the Navigator can only access the high-fidelity simulation generated by the *Somatic logos* (*Phenomena*). This structural gap is the origin of **C2 (The Confidence Gradient)**. See Immanuel Kant,

a **simulator**, running "what-if" scenarios and predictive models to test potential actions against the *Logos* before they are committed to.[15]

6.3.4 The Integrator: *Nous* as the Conscious Navigator

- **Mode:** Integrative / Decisional (*Prohairesis*).
- **Modern Correlate:** Global Workspace / Conscious Awareness[16]
- **Function:** The *Nous* is the apex of the system—the emergent space where integration occurs. It takes the raw sensory feed (*Aisthēsis*) and merges it with the analytical models presented by the *Logistikon*.
    - It creates the "synchronic snapshot" of the present (the *Somatic Present*).
    - It performs the final act of **Decision** (*Prohairesis*).
    - It is the only faculty capable of overriding the inertia of the unconscious (*Thymos* and *Orexis*) by directing the *Logistikon* to re-frame the situation.

The Metabolic Switch: Crucially, the *Nous* is not permanently "online" in its full, deliberative mode.

- **The Automated Dispatcher:** For the vast majority of existence, the *Logistikon* operates on **Auto-Pilot**. It functions as an efficient, low-energy dispatcher, automatically prioritizing and ordering the impulses of *Orexis* and *Thymos* based on established habits and heuristic success.[17]

Critique of Pure Reason, trans. Paul Guyer and Allen W. Wood (Cambridge: Cambridge University Press, 1998), A235/B294–A260/B315.

15. Views the mind as a generator of expectations and simulations. Daniel C. Dennett, *Kinds of Minds: Toward an Understanding of Consciousness* (New York: Basic Books, 1996), 96–112.

16. Neuroscientific correlate for the *Nous* as the space where information is integrated and broadcast. Bernard J. Baars, *A Cognitive Theory of Consciousness* (Cambridge: Cambridge University Press, 1988); Stanislas Dehaene, *Consciousness and the Brain: Deciphering How the Brain Codes Our Thoughts* (New York: Viking, 2014).

17. Explains the cognitive miserliness of the brain; we default to habit to save energy. Kahneman, *Thinking, Fast and Slow*, 35–48.

- **The Exception Handler:** The conscious *Nous* "kicks in" only when this automated ordering fails. This occurs when the *Logistikon* encounters **Novelty** (a situation with no precedent) or **Conflict** (where *Thymos* and *Orexis* are at an impasse that habit cannot resolve).[18]
- **The Evolutionary Logic:** Activating the *Nous* to generate new solutions is a metabolically expensive task. Nature reserves this high-energy state for moments where the standard operating procedure is insufficient to ensure survival or flourishing.[19]

In the unexamined life, the *Nous* remains dormant, activating only for survival emergencies, leaving the system driven by habituated *Doxa*. In the *Navigator*, the *Nous* cultivates the capacity to "switch on" at will, actively intervening to upgrade the *Logistikon's* automated protocols toward a higher harmony of *Dikaiosynē*.

"Consciousness is the gateway to the brain, a 'global workspace' that allows the error-prone, specialized functions of the mind to integrate, solve conflicts, and update their internal maps." — Bernard Baars[20]

---

18. Identifies the Anterior Cingulate Cortex function of conflict monitoring, paralleling the NPN view of *Nous* activation. Amitai Shenhav, Matthew M. Botvinick, and Jonathan D. Cohen, "The Expected Value of Control: An Integrative Theory of Anterior Cingulate Cortex Function," *Neuron* 79, no. 2 (2013): 217–40.

19. Biophysical justification for the rarity of high-level conscious thought; it is energetically costly. Marcus E. Raichle, "The Brain's Dark Energy," *Science* 314, no. 5803 (2006): 1249–50; David Attwell and Simon B. Laughlin, "An Energy Budget for Signaling in the Grey Matter of the Brain," *Journal of Cerebral Blood Flow & Metabolism* 21, no. 10 (2001): 1133–45.

20. Adapted from Bernard J. Baars, A Cognitive Theory of Consciousness (Cambridge: Cambridge University Press, 1988). Baars' Global Workspace Theory provides the neuroscientific correlate for the Nous as the emergent integrator. It explains why the system must shift from low-energy, automated processing (the Logistikon on auto-pilot) to high-energy conscious awareness: to resolve the "impasse" between competing drives and broadcast a new, aligned strategy to the whole.

Table 6.2 Topography of Navigation: Mapping Consciousness

Layer	Mode	Modern Correlate	Function
Orexis (Instinct)	Reactive / Hardwired	Autonomic / Limbic System (System 1)	The domain of instinct; manages biochemical needs (hunger, pain) below the threshold of awareness.
Thymos (Spirit)	Automated Heuristic	Social Intuition / "Hot" Cognition	The social calculator; scans for threats (*Neikos*) and alliances (*Philia*) to generate emotional conclusions.
Logistikon (Executive)	Analytical / Algorithmic	Executive Function (System 2)	The interface and simulator; translates "hot" signals into coherent models and tests "what-if" scenarios.
Nous (Navigator)	Integrative / Decisional	Global Workspace / Conscious Awareness	The emergent integrator; the only faculty capable of overriding inertia via conscious decision (*Prohairesis*).

6.4 The Constitutive Polarity of the *Psyche*: Cognitive *Philia* and *Neikos*

The cosmic forces of *Philia* and *Neikos* are not merely external dynamics to be navigated; they are the fundamental, constitutive forces of the *Psyche* itself. The mind is a microcosm of the *Archē*, and its operation is the perpetual interplay of these two drives:

- **Cognitive *Philia* (The Drive to Cohere):** This is the mind's impulse toward order, stability, and unity. It is the force that builds internal models, seeks logical consistency, creates habits, and strives for a unified, coherent worldview. It is the engine of understanding and the source of our desire for a stable, predictable reality.

- **Cognitive *Neikos* (The Drive to Differentiate):** This is the mind's impulse toward novelty, doubt, and deconstruction. It is the force that generates new ideas, questions assumptions, introduces dissonance, and breaks down outdated or flawed cognitive models. It is the engine of learning and the source of our capacity for adaptation and creativity.

The health of the *Psyche*—its *Dikaiosynē*—is not the victory of one force over the other, but their functional harmony.[21] A mind dominated by cognitive *Philia* becomes rigid and dogmatic, unable to adapt. A mind dominated by cognitive *Neikos* becomes chaotic and ungrounded, unable to act.

The *Logistikon's* primary ethical and epistemic task is to **orchestrate the internal dance of these forces.** It must know when to deploy cognitive *Neikos* to challenge a comfortable *Doxa*, and when to deploy cognitive *Philia* to solidify a hard-won *Epistêmê* into a reliable map for action.

It is from this primary, internal polarity that the secondary, methodological archetypes of the *Demiourgos* and *Genetor* emerge as recognizable patterns or "modes." The *Demiourgos* mode is a sustained expression of cognitive *Philia*; the *Genetor* mode, of cognitive *Neikos*. The masterful Navigator is not defined by a fixed identity in either mode, but by the practical wisdom (*Phronēsis* [Φρόνησις]) to consciously manage this internal polarity, applying the right mixture of cohesive and differentiating force demanded by the specific challenge within the *Logos*.

This analysis, of course, is itself a necessary flattening. The *Psyche* is a unified process, not a battleground of discrete forces. Cognitive *Philia* and *Neikos* are not two things that interact, but the two names we give to the inseparable poles of a single mental process—the inhale and exhale of thought. By the **Zero Principle**, this polarity is the transcendental condition for thought itself: for a concept or model to possess identity, it must emerge through the simultaneous force of differentiation (*Neikos*)

---

21. Highlights the necessity of emotional/intuitive inputs (Neikos/Philia forces) for balanced reasoning. Damasio, *Descartes' Error*, 165–201.

and coherence (*Philia*). We parse this unified flow into distinct forces for the same reason the *Somatic logos* creates all synchronic snapshots: to navigate. This map of the mind's polarity is perhaps the most essential one, but we must never mistake its contours for the living territory of consciousness itself.

Table 6.3   The Internal Polarity of the *Nous*

Feature	Cognitive Philia (Φιλία) (Drive to Cohere)	Cognitive Neikos (Νεῖκος) (Drive to Differentiate)
Core Impulse	Union, Order, Model Building, Stability.	Repulsion, Differentiation, Doubt, Novelty.
Goal	Unified, Coherent Worldview.	Adaptation, Learning, Paradigm Shift.
Manifestation	Logical Consistency, Habit, Trust, Loyalty.	Critical Analysis, Dissonance, Boundary Setting, New Ideas.
Pathology	Rigidity, Dogmatism, Inability to Adapt.	Chaos, Nihilism, Inability to Act or Sustain Focus.

6.5   Conclusion: The Grounded Navigator

With the full structure of the *Psyche* articulated, we have completed the first half of the Navigator's map. We have shown that the mind is not a spectral anomaly but a natural, emergent layering of the constitutive *Hormē*—the fundamental impulse to strive. This structure resolves the ancient split between body and mind, grounding agency in the very laws of the *Archē*.

Crucially, this stratification provides the **functional anchor for objective value** (C4). 'Good' is no longer a matter of opinion; it is simply the effective alignment of the striving *Psyche* with the external *Logos*. The navigator's goal, *Eudaimonia* (εὐδαιμονία), is thus revealed not as an abstract ideal, but as the demonstrable state of functional harmony (*Dikaiosynē*) achievable by a striving system within the lawful *Physis*.

CHAPTER 6. ARCHITECTURE OF THE NAVIGATOR

We now know *what* the navigator is, and *what* its objective purpose must be. But to achieve that purpose, the navigator must be able to read its own instruments.

The next stage of inquiry must turn inward. We must now ask: How does this emergent mind actually process the world? And, perhaps more importantly, what are the inherent limits and biases of its own tools of reason?

Table 6.4 The Epistemic Chain: NPN Terms for Chapter 7

NPN Term	Role	Technical Definition
Aisthesis (Αἴσθησις)	Perception	The foundational stage of the epistemic process; the direct, causal interaction between the *Archē* and the nervous system, providing raw sensory data.
Empeiros (Ἔμπειρος)	Experience	The second stage; accumulated experience and pattern recognition formed through memory and repetition from *Aisthēsis*.
Epistēmē (Ἐπιστήμη)	Knowledge	Justified, generic knowledge; a high-confidence model of the *Logos* validated by predictive success.

Chapter 7

NPN Epistemology: The Somatic *logos* and the Confidence Gradient

"The intellect is a mere instrument—a means to the furtherance of life; and its test is not how much truth it knows, but how much life it enables one to live." — Hans Vaihinger[1]

7.1 Introduction: The Somatic *logos* and the Confidence Gradient

The Stratified *Psyche* has shown us the architecture of the knower, and the *Hormē* has defined its objective mission. Yet, the history of philosophy, from Plato's Forms to Descartes' skepticism, is a catalogue of errors rooted in a flawed premise: that the mind's internal logic is a perfect mirror of the external cosmos.

Neo-Pre-Platonic Naturalism rejects this view. The mind's operating system—its **Somatic *logos*** —is not a transcendent gift but a **highly refined evolutionary heuristic**. It is a cognitive compression algorithm, honed over eons for survival in the macroscopic environment of the *Archē*.

This chapter undertakes the necessary task of diagnosing the Navigator's tools. We will show that the laws of logic are not eternal Platonic truths,

---

1. Hans Vaihinger, *The Philosophy of 'As If': A System of the Theoretical, Practical and Religious Fictions of Mankind*, trans. C. K. Ogden (London: Kegan Paul, Trench, Trubner & Co., 1924), viii. This establishes the NPN position of *Heuristic Realism*: cognitive models are judged by their adaptive utility (survival), not their isomorphic correspondence to an absolute reality.

but **adaptations** that lead to an inescapable **synchronic bias** (C3)—the functional illusion of the "now."

By fully characterizing the *Somatic logos* and tracing the epistemic process from raw perception (*Aisthēsis*) to justified model (*Epistēmē*), we establish the true measure of knowledge: the **Confidence Gradient** (C2). This diagnosis is essential. Only by understanding the fallibility and inherent bias of its own reason can the *Nous* truly be set free to navigate the objective *Logos* of the cosmos.

7.1.1 The Parmenidean Leap and Its Category Error

Parmenides made a monumental discovery: the Somatic *logos* of the mind possesses a compelling, *a priori* necessity. His category error was to **reify** this model, concluding that reality *itself* must conform to its rules, thus denying change, multiplicity, and non-being. He correctly mapped the *Nous*'s operating system but erroneously assumed the external *Archē* must run on the same software. This inaugurated a perennial philosophical error: mistaking the structure of the navigator for the structure of the territory.

7.2 The Somatic *logos* as an Evolved Heuristic

The NPN resolution is that the Somatic *logos* is not a glimpse of the cosmos's fundamental fabric but a **highly refined evolutionary heuristic**. The minds that survived were those that modeled the world as consisting of discrete, self-identical objects persisting through time and obeying consistent cause-and-effect.[2] The laws of logic are reliable not because they mirror a transcendent realm, but because they are adaptations for effective action in the macroscopic world where our cognition evolved.[3] They

---

2. This aligns with the central thesis of Embodied Cognition, which argues that abstract reason is metaphorically grounded in sensorimotor experience. See George Lakoff and Mark Johnson, *Philosophy in the Flesh: The Embodied Mind and Its Challenge to Western Thought* (New York: Basic Books, 1999), 3–20, 77–105. NPN radicalizes this: logic *is* the body's physics engine.

3. See Daniel C. Dennett, *From Bacteria to Bach and Back* (New York: W. W. Norton & Company, 2017); and Patricia Churchland, *Touching a Nerve: The Self as Brain* (New York: W. W. Norton & Company, 2013) for the evolutionary and neurobiological basis of reason.

are the **Somatic logos**—the logic of the body—a navigation system honed for a specific, survival-relevant slice of the *Archē*.[4]

7.2.1 Bridging the Gap: Neural Architecture as a Mirror of Causal Structure

This is not a retreat to mere pragmatism. The reason the Somatic *logos* is so effective is that the neural architecture of the brain *itself* evolved within and was shaped by the causal structure of the external *Logos*. The brain is a physical system that must obey the very laws of physics it is trying to model. Its wiring, its learning algorithms, and its resulting logical axioms are a reflection of the environment that built it. The feeling of *a priori* necessity is the feeling of using a tool that is so deeply embedded in our cognitive apparatus that we cannot consciously perceive its construction. It is the **Somatic logos**—the logic of the body (*soma*, Σῶμα)—a reasoning capacity inextricably tied to our physical, evolutionary history.[5]

> **First Principle**
>
> **FP7: FIRST PRINCIPLE OF EPISTEMIC GROUNDING: THE SOMATIC *LOGOS***
>
> **Statement:** The operational logic of the evolved *Nous* is functionally aligned with the operational *Logos* of the *Archē* because the former is a product of the latter.
>
> **Justification:** The negation is a performative contradiction. A *Nous* fundamentally misaligned with the *Logos* would produce maladaptive models, leading to extinction long before it could develop the capacity for abstract inquiry. To deny this principle is to use a functioning navigator to argue for the impossibility of navigation.

---

4. For the philosophical framework of reason as a somatic adaptation tied to experience, see John Dewey, *Experience and Nature* (Chicago: Open Court, 1925), chap. 6.

5. See Dewey, *Experience and Nature*, chap. 6. By grounding logic in the *Soma* (body), NPN bypasses the Rationalist/Empiricist deadlock. Reason is "empirical" because the body is an evolved record of past experience.

7.2.2 The Somatic *logos* and the Synchronic Snapshot

The *Logistikon* operates according to a seemingly self-evident internal logic. The history of philosophy, however, is in many ways the history of misinterpreting the origin and scope of this logic. The NPN resolution is that the Somatic *logos* is not a glimpse of the fundamental fabric of the cosmos, but a highly refined evolutionary heuristic.

This heuristic has a specific and consequential operational mode: it generates **synchronic snapshots**. To navigate a dynamic, four-dimensional *Physis*, the *Nous* must create static, three-dimensional models—discrete objects, binary states, and linear cause-and-effect sequences. This is the source of its power and its fundamental bias. The thought "do this or that" is the micro-instance of this process: a complex, diachronic reality is flattened into a binary choice between two frozen potential futures. This is not a flaw but a feature—the necessary method of a finite navigator. The error occurs when this navigational tool is mistaken for a perfect reflection of reality itself, a confusion that lies at the root of the Parmenidean flattening and its philosophical consequences.

7.2.3 The Tyranny of the "Now": The Somatic Root of Synchronic Bias

The *Somatic logos* did not evolve to model reality perfectly, but to facilitate survival through decisive action. This imperative reveals the deepest source of our synchronic bias: **the "now" is the non-negotiable point of action.**

Consciousness operates in a perpetual, felt present. To make a decision—to eat, to flee, to speak—the vast, diachronic complexity of cause, effect, and potential futures must be collapsed into a single, actionable moment.[6]

---

6. The analysis of the Synchronic Snapshot aligns with Henri Bergson's critique of the "cinematographic mechanism" of the intellect. Bergson argued that the intellect misrepresents true time (*Durée*) by spatializing it into discrete, sequential instants. See Henri Bergson, *Duration and Simultaneity: Bergson and the Einsteinian Universe* (Manchester, UK: Clinamen Press, 1998), 44–50.

CHAPTER 7. NPN EPISTEMOLOGY: THE SOMATIC LOGOS 83

The feeling of a binary choice ("do this or that") is the cognitive compression of a four-dimensional probability landscape into a synchronic snapshot. This snapshot is the mind's **user interface** for the *Physis*; it is a necessary and functional illusion.

The Diachronic Turn, therefore, is not the natural state of the operating system, but a learned **meta-cognitive skill**—the ability to see the single frame *as a frame* and consciously reconstruct the dynamic movie of which it is a part.

> **Corollary**
>
> **C3: COROLLARY: THE SOMATIC PRESENT AS A NAVIGATIONAL SNAPSHOT**
>
> **Statement:** The conscious experience of a synchronic "now" is a computational necessity for any finite Navigator. It is the operational interface that collapses the diachronic complexity of the *Archē* into an actionable moment.
>
> **Justification:** This follows directly from the First Principle of Becoming: Diachronic Primacy. Because the *Archē* is a continuous flux, it presents an infinite stream of causal data. A finite Navigator cannot process infinite continuity in real-time. Therefore, to calculate a decision, the *Nous* must perform a **dimensional reduction**—freezing the 4D process into a static 3D model (the snapshot). The negation is a mathematical absurdity: to claim a finite agent can act without a snapshot is to assert it can calculate a vector of action without a coordinate of origin. Since a dynamic flux offers no inherent fixed points, the snapshot is the non-negotiable zero-point required for the computation of agency.

The inherent bias toward the synchronic snapshot is not merely a philosophical misstep; it is the **somatic root** of the error itself. The very structure of our consciousness—the felt "now" in which all decisions are made—is a biological compression algorithm. It forces the diachronic, four-dimensional reality of the *Physis* into a series of actionable, but fundamentally flattened, moments. **The entire history of synchronic philosophy can thus be understood as the intellect being misled by its own user interface, mistaking the necessary heuristic of the present for the fundamental nature of reality.**

> ### Theorem
>
> #### T3: THEOREM: THE NECESSARY DISTORTION
>
> **Statement:** Every synchronic model generated by the *Nous* is strictly ontologically distinct from and descriptively incomplete relative to the diachronic *Archē* it represents. Therefore, "distortion" is not a failure of cognition, but a structural requirement for navigation.
>
> **Derivation:** This theorem describes the inevitable gap between the map and the territory.
>
> 1. The *Archē* is a continuous, unfolding process of Becoming (**FP2: Diachronic Primacy**).
> 2. The *Navigator* must perform a dimensional reduction, freezing the 4D process into a static 3D snapshot to make a decision (**C3: The Somatic Present**).
> 3. A static snapshot (t_0) cannot contain the vector of change ($t_0 \to t_1$) without converting it into a static symbol.
> 4. Therefore, the model must fundamentally differ in nature from the reality it models to be usable.
>
> **Conclusion:** Just as a photograph is a distortion of movement, a concept is a distortion of process. The *Navigator* must trade *fidelity* (completeness) for *utility* (actionability).

7.3 The Epistemic Process: From Perception to Justified Knowledge

Neo-Pre-Platonic Naturalism posits a continuous, iterative process through which the *Nous* converts raw sensory data into actionable, justified knowledge. This process, which moves from the particular to the generic, is the mechanism by which the **not-yet-known *Archē*** is systematically transformed into the **known *Archē***, a perpetual navigation bounded by the silent *Apeiron*.

7.3.1 *Aisthēsis* (Αἴσθησις): Sensory Data Acquisition

The epistemic chain begins with *Aisthēsis*—sensory perception. This is the direct, causal interaction between the *Archē* and the nervous system, providing a continuous stream of data about the immediate environment. In NPN, *Aisthēsis* is not considered infallible, but it is the non-negotiable starting point. It is the source of all empirical content, the raw material upon which the *Nous* operates. Without this anchor in the physical *Archē*, the Somatic *logos* would have nothing to model, leading to the sterile, self-referential systems Parmenides fell into.[7]

7.3.2 *Empeiros* (Ἔμπειρος): Pattern Recognition and Associative Learning

The second stage is *Empeiros*—accumulated experience. Through memory and repetition, the *Nous* identifies statistical regularities and causal sequences within the stream of *Aisthēsis*. This is the formation of heuristic patterns: "fire is hot," "predators stalk," "certain berries cause illness." *Empeiros* is practical, context-dependent, and often non-verbal. It represents the mind's first successful compressions of the *Logos* into efficient, if sometimes flawed, rules of thumb. It is the foundation of all learned behavior.

---

7. For Parmenides' privileging of reason (*logos*) over the senses (*aisthēsis*), see Fragment B7 (DK 28) in G. S. Kirk, J. E. Raven, and M. Schofield, *The Presocratic Philosophers: A Critical History with a Selection of Texts*, 2nd ed. (Cambridge: Cambridge University Press, 1983), 248–50. The fragment cautions against relying on habitual sensory impressions ("aimless eye or ear and tongue full of meaningless sound") in favor of rational judgment.

7.3.3 *Epistēmē* (Ἐπιστήμη): The Formation of Testable, Causal Models

The culmination of the process is ***Epistēmē***—justified, generic knowledge. This is not merely a collection of experiences, but a structured, abstract model that explains the *why* behind the patterns of *Empeiros*. *Epistēmē* is characterized by:

- **Genericity:** It applies beyond a specific instance (not "this fire is hot," but "combustion releases kinetic energy").
- **Causal Explanation:** It posits underlying mechanisms and relationships governed by the *Logos*.
- **Justification:** Its validity is supported by logical coherence and, crucially, by its continued predictive success when tested against new data from *Aisthēsis*.

Epistēmē is the functional understanding of the *Logos* that enables the *Nous* to plan and act with long-term efficacy. Crucially, the entire process is a **feedback loop**: *Epistēmē* informs what to look for in *Aisthēsis*, which refines *Empeiros*, which in turn tests and updates *Epistēmē*.

It is therefore essential to recognize that *Epistēmē* is not a state of certain, final possession of truth. The model created by the *Nous* is a map, and the map can never be the territory (the *Logos* of the *Archē* itself). This gap is unbridgeable. Consequently, *Epistēmē* is more accurately described as a **Confidence Gradient** rather than a binary state. It represents a model of such high functional utility and predictive success that it is rationally warranted, but it remains forever provisional. This inherent uncertainty is not a flaw but a fundamental feature of an emergent *Nous* operating within and upon *Physis*.

If the *Nous* is an evolved *navigator*, then how does it actually revise its maps when they fail? The process is formalized in the Socratic method, which NPN identifies as a cognitive algorithm for crisis and renewal.

Table 7.1 The Epistemic Chain: From Data to Model

Stage	Input	Process	Output
Perception (*Aisthēsis*)	Raw physical data from the *Archē*.	Causal interaction with the nervous system.	**Data:** Immediate sensory impression.
Experience (*Empeiros*)	Memory of repeated *Aisthēsis*.	Pattern recognition and associative learning.	**Heuristic:** Practical "rule of thumb" (e.g., Fire burns).
Knowledge (*Epistēmē*)	Aggregated *Empeiros*.	Causal modeling and predictive testing.	**Model:** Abstract, justified theory (e.g., Thermodynamics).

7.4 Primary Corollary: The Confidence Gradient of Epistemic Justification

"The gods did not reveal all things to mortals from the beginning; but by seeking as time goes on, they find out better."

— Xenophanes of Colophon[8]

The epistemic process, from *Aisthēsis* to *Empeiros* to *Epistēmē*, describes *how* a finite *Nous* builds knowledge. The outcome of this process—a high-confidence, yet provisional model—is not a compromise or a failure to achieve a higher ideal.[9] It is the *only logically possible form* that justified knowledge can take for an emergent navigator within the *Physis*.

To demand that *Epistēmē* must be certain, static, and final is to misunderstand the fundamental relationship between the knower and the known.

---

8. Xenophanes, DK 21 B18. This fragment provides the historical basis for C2 (The Confidence Gradient). It shifts the goal of the Navigator from the sudden "revelation" of absolute truth to the gradual, diachronic refinement of models over time.

9. This continuous process of perceptual inference and model-updating is formalized in Friston's Free-Energy Principle as the minimization of prediction error. See Karl Friston, "The Free-Energy Principle: A Unified Brain Theory?" *Nature Reviews Neuroscience* 11, no. 2 (2010): 127–38. This biological imperative establishes the *Confidence Gradient*: models are sustained by their ability to minimize surprise (entropy), not by their absolute truth.

The *Nous* is not a transcendent spectator; it is a finite subsystem, generated *within* the *Archē*, attempting to model the very system that contains it. The map is a part of the territory it represents.

This leads to an inescapable conclusion about the nature of knowledge itself.

7.4.1 The Domain of Certainty: A Vital Distinction

It is crucial here to distinguish between the *tools* of navigation and the *act* of navigation. The "Confidence Gradient" applies specifically to **Synthetic Knowledge**—claims about the structure, behavior, and content of the physical *Archē*.

Analytic Truths—such as the axioms of mathematics, formal logic, and the First Principles of this system itself—retain a form of **formal certainty**. Their validity is derived from internal consistency and definition, not empirical observation. $1 + 1 = 2$ is absolutely true because it is a closed logical loop. However, the moment such a formal truth is *applied* to model the *Archē* (e.g., "one apple plus one apple"), it constitutes a synthetic model subject to the constraints of the *Apeiron* and is evaluated on the Confidence Gradient.

We can be certain about the *grammar* of the *Logos* (our logical tools), but we must remain provisional about our *reading* of the text (our specific models of reality). The Navigator trusts the compass (Logic) implicitly, even while acknowledging that the chart of the waters (Epistēmē) is always being refined.

This is the vital mistake that has fundamentally divided idealism from materialism. Idealists mistake the certainty of the internal, analytic tool for the certainty of the external, synthetic world. Materialists, by dismissing the analytic *Nous* as a mere byproduct, lose the ability to ground objective, formal certainty entirely. NPN resolves this schism by establishing the proper relationship: Analytic Truths provide the unshakeable grammar of the mind; Synthetic Knowledge provides the provisional content of reality.

CHAPTER 7. NPN EPISTEMOLOGY: THE SOMATIC LOGOS

> **Corollary**
>
> **C2: PRIMARY COROLLARY: THE CONFIDENCE GRADIENT OF EPISTEMIC JUSTIFICATION**
>
> **Statement:** *Epistêmê* is a provisional, high-fidelity model of the *Logos*, justified by its predictive success and functional utility in navigating the *Archē*. Its measure is a *Confidence Gradient*, not a binary state of certainty.
>
> **Justification:** This is the direct epistemological consequence of the *First Principle of Diachronic Primacy*. A finite model generated *within* a system cannot contain a perfect, complete representation of that system. The negation—demanding certain, final knowledge—is a logical impossibility, as it requires that a finite *Nous*, itself a small subset of the *Archē*, could generate a synchronic snapshot of the *Archē* that is certain, complete, and final. The Confidence Gradient is therefore the *only possible form* that justified knowledge can take for a finite, emergent navigator.

7.5 Conclusion: The Grounded Navigator

This chapter has completed a fundamental synthesis, moving from a metaphysics of a dynamic *Archē* to a coherent, emergentist account of the *Nous* that inhabits it. The result is a robust naturalism where the knower is no longer a spectral exception but is fully integrated into the known.

We have shown that the stratified *Psyche* is not a random assemblage but an evolutionary layering of the core *Hormē*, where each stratum—*Orexis, Thymos, Logistikon, Nous*—functions as a sophisticated instrument for navigating a different facet of the *Logos*.

We have, therefore, achieved three critical goals:

1. **We have grounded agency in the *Archē*,** deriving the *Hormē* as the constitutive, non-negotiable ground of any striving system.

2. **We have grounded the mind in agency,** showing the **quadripartite, emergent structure** of the *Psyche* to be a natural consequence of a navigating *Hormē* within the *Archē's* dynamics.
3. **We have grounded value in the mind,** deriving objective "good" from the functional alignment of a striving system with the lawful *Logos*.

The *navigator* is no longer a philosophical abstraction. It is a physical system, endowed with a specific quadripartite architecture and driven by a constitutive impulse to persist. We now know *what* the navigator is, and *what* its objective purpose must be (*Eudaimonia*).

But to achieve that purpose, the navigator must be able to read its own instruments. The *Psyche* may be structured to navigate, but its tools—reason, logic, and perception—are not infallible.

Table 7.2 The Navigator's Path: NPN Terms for Chapter 8

NPN Term	**Role**	**Technical Definition**
Eukairia (Εὐκαιρία)	Receptivity	The precondition of cognitive receptivity; the moment when the *Psyche* is psychologically ready to entertain model flaw.
Exaiphnes (Ἐξαίφνης)	Rupture	The sudden, jarring metacognitive awareness of a model's failure that accompanies the *Lysis*.
Energeia (Ἐνέργεια)	Actuality	"Being-at-Work." The active, sustained expression of a *Nous* that has successfully realigned itself with the *Logos*.
Prohairesis (Προαίρεσις)	Choice	The fundamental methodological commitment between the *Demiourgos* and *Genetor* modes of engagement.
Phronēsis (Φρόνησις)	Wisdom	Practical Wisdom; the meta-cognitive capacity to discern which mode (*Demiourgos* or *Genetor*) is required by the specific context.

CHAPTER 7. NPN EPISTEMOLOGY: THE SOMATIC LOGOS

NPN Term	Role	Technical Definition
Eudaimonia (εὐδαιμονία)	Flourishing	The objective flourishing of a life well-navigated; the successful outcome produced by a *Psyche* in a state of *Dikaiosynē*.
Demiourgos (Δημιουργός)	Mode A	The Master Craftsman mode; seeks mastery through structural analysis, order, and replication.
Genetor (Γενέτωρ)	Mode B	The Begetter mode; seeks creation through synthesis, novelty, and the generation of new models.

The final inquiry must therefore turn inward before we can act. **How does this emergent mind actually process the world?** And are the maps it draws—its "logic" and "truths"—accurate reflections of reality, or merely useful evolutionary illusions? To answer this, we must examine the *Somatic logos*.

Chapter 8

The Art of Navigation: From Doxa to Energeia

"The unexamined life is not worth living for a human being." — Socrates[1]

8.1 Introduction: The Practice of Philosophical Realignment

The theoretical edifice is complete. We have systematically derived the First Principles of Neo-Pre-Platonic Naturalism, establishing a coherent framework that spans metaphysics, epistemology, and psychology. We possess a map of the dynamic *Archē* and a detailed schematic of the emergent Navigator—the stratified *Psyche* with its core *Hormē*, its *Somatic logos*, and its capacity for meta-cognitive choice.

But a map is not a journey. A schematic is not a lived experience. The ultimate purpose of this entire philosophical recovery mission is not merely to understand the world, but to *Navigate* it effectively—to transition from a state of unexamined *Doxa*, where the *Psyche* is misaligned with the *Logos*, to a state of active *Energeia*, where the *Nous* is fully at-work in the world.

This chapter is the guide to that transformation. It synthesizes our first principles into a **methodology for cognitive and ethical realignment**.

---

1. Plato, *Apology*, 38a. NPN interprets "examination" not as moralizing, but as the rigorous auditing of the *Psyche's* navigational maps against the *Logos*. In *Plato: Complete Works*, ed. John M. Cooper (Indianapolis, IN: Hackett Publishing Company, 1997), 33.

We will move from theory to practice, outlining the concrete stages of the navigational process:

- The diagnosis of cognitive sickness (*Doxa*) and the induction of therapeutic crisis (*Elenchus, Lysis*).
- The navigation of the resulting void (*Aporia*) and the re-anchoring in reality's constraints (*Dikē*).
- The conscious choice of navigational mode (*Prohairesis*) and the ultimate expression of a realigned life: active flourishing (*Energeia* and *Eudaimonia*).

This is the art of navigation. It is the applied science of aligning the Somatic *logos* of the soul with the external *Logos* of the cosmos. The following sections provide the manual for this most vital of journeys.[2]

8.2 The Induction of Cognitive Dissonance: The Socratic Method

The epistemic process described thus far operates within an accepted paradigm. However, the growth of knowledge often requires the failure and replacement of a foundational model. Neo-Pre-Platonic Naturalism formalizes this transformative crisis using the Socratic method as its archetypal template—a deliberate, structured induction of cognitive dissonance to force a system-level upgrade and realign the *Hormē* with the *Logos*.[3]

8.2.1 *Eukairia* (Εὐκαιρία): The Precondition of Cognitive Receptivity

The process cannot begin without *Eukairia*—the receptive moment. This is a specific configuration of the *Psyche* where the *Thymos* is not so inflamed by defensiveness that it blocks inquiry, and the *Logistikon* is active

---

2. This conception of philosophy as a practical art of living—an alignment of the self with cosmic order—is central to Pierre Hadot's interpretation. NPN adopts this therapeutic model: theory is useless unless it culminates in the *Energeia* of the Navigator. See Pierre Hadot, *Philosophy as a Way of Life: Spiritual Exercises from Socrates to Foucault*, trans. Michael Chase (Oxford: Blackwell, 1995), 81–125.

3. For the classic demonstration of this method, see Plato, *Meno*, 80a–86c. NPN identifies Meno's paradox as the moment of *System Failure* required to initiate a higher-order learning process.

and curious. It is a state of psychological readiness, a teachable moment where the individual is capable of entertaining the possibility that their current model—*Doxa*—may be flawed. Without this fertile ground, the seed of doubt cannot take root.[4]

8.2.2 *Elenchus* (Ἔλεγχος): The Logical Interrogation of Belief Structures

Elenchus—cross-examination—is the systematic application of the Somatic *logos* to test the structural integrity of a belief.[5] Beginning from an individual's own stated position (whether expressed outwardly or held internally), a series of precise, logical questions guides them to expose internal contradictions, unwarranted assumptions, and inconsistencies within their model.

This is not merely an external dialogical tool but can be—and in the mature Navigator must be—conducted as an internal stress test. The *Elenchus* probes the existing cognitive structure, identifying points of strain until a critical weakness is found.[6]

8.2.3 *Lysis* (Λύσις) & *Exaiphnes* (Ἐξαίφνης): Model Dissolution and Metacognitive Awareness

The discovery of a fatal contradiction within the model triggers the *Lysis*—the cognitive dissolution, where the foundational assumptions of the belief structure are broken down.[7] This structural failure is immediately accompanied by the *Exaiphnes*—the sudden, jarring metacognitive awareness of the failure. It is the "Aha!" moment of profound disorientation,

---

4. Cf. Jonathan Haidt, *The Righteous Mind: Why Good People Are Divided by Politics and Religion* (New York: Pantheon Books, 2012). Haidt's "elephant" corresponds to the NPN *Thymos*; without *Eukairia* (calming the elephant), the "rider" (*Logistikon*) cannot steer.

5. For Gregory Vlastos's seminal analysis of the Socratic elenchus, see "The Socratic Elenchus," *Oxford Studies in Ancient Philosophy* 1 (1983): 27–58. NPN reframes this not just as logical checking, but as a *structural stress test* for the Navigator's models.

6. This productive holding of cognitive tension aligns with Alicia Juarrero's dynamic systems approach. See Alicia Juarrero, *Dynamics in Action: Intentional Behavior as a Complex System* (Cambridge, MA: MIT Press, 1999), 131–50. NPN argues that intentionality emerges from the *Nous's* ability to constrain and direct this tension.

7. For an example of *Elenchus* in action, see Plato, *Euthyphro*, 7a–11b. Euthyphro's failure to define piety is the *Lysis* of his unexamined confidence.

where the individual realizes, "What I thought I knew, I do not know." This is the rupture in the previously stable epistemic ground.[8]

8.2.4 *Katharsis* (Κάθαρσις): The Post-Collapse Purge of *Doxa*

The *Lysis* breaks the flawed model; *Katharsis* is the conscious, often difficult choice to purge its influence from thought and action. It is the moment where the individual, having seen the contradiction, must decide: to realign behavior with the new understanding, or to cling to the comfort of the old *Doxa* despite knowing it is false. This is not merely emotional release, but ethical and cognitive housecleaning—the removal of contradictions from one's operational worldview. Without *Katharsis*, the *Elenchus* is merely an intellectual exercise; with it, the *Psyche* begins the practical work of realignment.[9]

8.3 Navigating the *Apeiron* (Ἄπειρον): The Post-Collapse Phase

The successful *Elenchus* does not end with the collapse of a belief; it initiates a critical and fertile period of disorientation. This phase is not a failure of the process, but its essential, purgative core—the necessary clearing of the cognitive ground.[10]

8.3.1 *Aporia* (Ἀπορία): The State of Problem Recognition and Pathlessness

The immediate consequence of the *Katharsis* is *Aporia*—a state of being at a loss, without a path (*poros*).[11] This is more than simple confusion; it is a productive and profound perplexity. The old paths of thought and

---

8. Cf. Thomas S. Kuhn, *The Structure of Scientific Revolutions*, 2nd ed. (Chicago: University of Chicago Press, 1970). The *Exaiphnes* is the micro-cosmic equivalent of a Kuhnian "crisis phase" preceding a paradigm shift.

9. For the foundational psychological work on insight, see Karl Duncker, "On Problem-Solving," *Psychological Monographs* 58, no. 5 (1945). *Katharsis* is the behavioral commitment that must follow the cognitive insight.

10. Plato, *Meno*, 80a–b.

11. For the epistemic role of aporia, see Gavin Lawrence, "The Aporia of the Meno," in *Maiensis* (Oxford: Oxford University Press, 2007). NPN validates *Aporia* as the "Zero Point" of inquiry—the necessary clearing of the *Apeiron*.

action, which were once taken for granted, are now closed.[12] The individual consciously holds a question without an answer, a problem without a solution. In the context of NPN, *Aporia* is the cognitive and experiential registration of the *Apeiron*. It is the mind, stripped of its flawed map, honestly confronting the fact that it is lost. This state of conscious ignorance is the only proper foundation for seeking true *Epistēmē*.[13]

8.3.2 Mapping the Unknown: Scoping the Boundaries of Current Ignorance

Following the acknowledgment of *Aporia*, a more analytical process begins. The individual moves from the sheer experience of being lost to the task of mapping the unknown. This involves scoping the boundaries of the newly revealed *Apeiron*: What are the precise contours of the problem? What assumptions have been invalidated? What new questions must now be asked? This is not yet problem-solving, but problem-framing.[14] It is the deliberate and sober assessment of the scale of one's ignorance, transforming the vague anxiety of *Aporia* into a defined set of challenges to be overcome. This act of mapping is the first step toward constructing a new path through reality itself.

8.4 *Dikē* (Δίκη): Acceptance of Causal Constraints

Following the disorientation of *Aporia* and the mapping of the *Apeiron*, the mind must re-anchor itself before reconstruction can begin. This re-anchoring is not found in a new dogma, but in a sober acceptance of the non-negotiable framework of reality itself—a phase termed *Dikē*—justice.[15]

---

12. Navigating the aftermath of cognitive collapse resonates with Martha Nussbaum's exploration of fragility. See *The Fragility of Goodness* (Cambridge: Cambridge University Press, 1986), 1–21. The *Navigator* must accept vulnerability to rebuild robustly.

13. Plato, *Theaetetus*, 155c–d. Wonder (*Thauma*) is the emotional correlate of *Aporia*—the acknowledgment of the gap between map and territory.

14. This aligns with the critical first phase of problem-solving identified in modern cognitive science. Cf. J. R. Hayes, *The Complete Problem Solver*, 2nd ed. (Hillsdale, NJ: Erlbaum, 1989).

15. The concept of *Dikē* as the acceptance of cosmic constraints is rooted in Anaximander. See Charles H. Kahn, *Anaximander and the Origins of Greek Cosmology* (New York:

8.4.1 Re-anchoring in the Non-Negotiable Parameters of the *Logos* (Λόγος)

Dikē represents the understanding and acceptance of the impartial, causal constraints of the *Archē*. After the collapse of a personal, subjective model (*Doxa*), the mind confronts the objective "justice" of the *Logos*—the impersonal laws of cause and effect that govern outcomes regardless of one's beliefs or desires. This is the realization that the problem, as defined by the newly mapped boundaries of the *Apeiron*, operates according to a specific, discoverable set of rules. *Dikē* is the commitment to playing by reality's rules, as dictated by the *Logos*, rather than attempting to impose one's own.

8.4.2 The Foundation for Model-Building: Acknowledging Empirical and Logical Boundaries

This acceptance forms the only solid foundation for building a new, more robust model. The parameters of the new model are now defined by two masters:

- **Empirical Boundaries:** The model must account for all relevant data from *Aisthesis*, including the anomalous data that triggered the original *Lysis*.
- **Logical Boundaries:** The model must be internally consistent, adhering to the *Somatic logos* to avoid the same pitfalls of contradiction that doomed the previous model.

Dikē is the moment of humility that follows the humbling of *Aporia*. It is the conscious decision to subordinate the *Psyche* to the external *Logos*, ensuring that any new structure of knowledge is built upon the bedrock of reality, not the sands of wishful thinking (*Doxa*).

Columbia University Press, 1960), 166–83. NPN secularizes this: justice is alignment with the causal *Logos*.

8.5 The Methodological Choice: *Prohairesis* (Προαίρεσις) in Inquiry

The state of *Dikē* (Δίκη) establishes what the problem is and the rules that govern it. The subsequent step, *Prohairesis*, is a deliberate commitment to *how* the problem will be engaged.[16] This is a fork in the path of inquiry, representing two fundamental orientations of the *Nous* (Νοῦς) toward the *Logos* (Λόγος) and the *Apeiron* (Ἄπειρον).

8.5.1 Mode A: Structural Analysis: The *Demiourgos* (Δημιουργός) Mode

This mode is characterized by the *Demiourgos*—the Craftsman archetype.[17] Its drive is to master reality through decomposition, analysis, and the perfect execution of a known blueprint or law.

- **Telos:** Prediction and Control.
- **Paradigmatic Forms:** The experimental method in science, analytical philosophy, reverse-engineering, and diagnostic reasoning.
- **Function:** To create and maintain order, stability, and high-fidelity replication.
- **Pathology:** An imbalance in this mode leads to stagnation and rigidity. A society or mind that is purely *Demiourgos* is a sterile prison, incapable of adaptation.

8.5.2 Mode B: Synthesis: The *Genetor* (Γενέτωρ) Mode

This mode is characterized by the *Genetor*—the Begetter archetype. Its drive is to generate novelty through combination, synthesis, and the creation of new models and forms not present in the existing blueprint.

---

16. For the classical analysis of *prohairesis*, see Aristotle, *Nicomachean Ethics*, 1112a15–1113a14. NPN elevates this to the *Meta-Cognitive Switch*: the choice of *how* to process reality.

17. The archetypal distinction matches M. J. Kirton, "Adaptors and Innovators," *Journal of Applied Psychology* 61, no. 5 (1976). NPN maps "Adaptors" to *Demiourgos* and "Innovators" to *Genetor*.

- **Telos:** Generation and Creation.
- **Paradigmatic Forms:** Theoretical model-building in physics, systems thinking, architectural design, entrepreneurship, and artistic creation.
- **Function:** To provide adaptation, novelty, and the transcendence of current paradigms.
- **Pathology:** An imbalance in this mode leads to fragmentation and incoherence. A society or mind operating purely as *Genetor* dissolves into chaos, lacking the stability to sustain its creations.[18]

Table 8.1   The Methodological Paradox: Tool vs. Goal

Archetype	Method (The Tool)	Mechanism	Telos (The Result)
Mode A (*Demiourgos*)	**Analysis** (Separation)	Decomposes complex reality to match a Blueprint.	**Order** (Social *Philia*)
Mode B (*Genetor*)	**Synthesis** (Union)	Combines disparate elements to create a New Model.	**Novelty** (Social *Neikos*)

8.6   *Energeia* (Ενέργεια): The Active Implementation of a New Model

The journey of conceptual revision culminates not in a static state of possession, but in a mode of being. *Energeia*—being-at-work, describes the active, sustained, and generative expression of a *Nous* that has successfully realigned itself and the *Hormē* it serves with the *Logos* through the preceding stages of crisis and choice.

---

18. Note on the Dialectic: A synchronic snapshot often mistakes the method for the goal. The Demiourgos uses the blade of analysis (separation) to carve perfect unity. The Genetor uses the bond of synthesis (union) to birth something unique and different. The master Navigator understands that these are not contradictions, but the fluid interplay of the Archē itself—a dynamic where, as in the ancient Yin-Yang polarity, each force contains the seed of the other to achieve its highest expression.

8.6.1 Cognitive Realignment as an Ongoing Process

Energeia is not the end of inquiry but its transformation. The new model is not treated as a final dogma. Instead, it is deployed as a **living, working hypothesis**. The *Logistikon* remains actively engaged in a feedback loop with the *Archē*, continuously testing the model's predictions against *Aisthesis* and refining it through *Empeiros*. The state of *Aporia* is never fully left behind; it is integrated as a permanent potentiality, a humble awareness that the current model is provisional and must be held open to future *Lysis* and revision.[19]

8.6.2 The Navigator's *Phronēsis*: Mastery of the Modes

It is a critical misinterpretation to view these two modes as a moral dichotomy of "good" vs. "evil". They are amoral, functional archetypes—descriptive modes of operation, not prescriptive judgments of value. Both are essential. The *Demiourgos* stabilizes the leaps of the *Genetor*, creating a new foundation for the next generative leap.

The Hierarchy of Agency This distinction reveals a fundamental hierarchy within the NPN model of mind:

1. **The Base State:** The unexamined individual operating on *Doxa* (Δόξα).
2. **The Navigator:** The individual whose *Nous* (Νοῦς) has attained the meta-cognitive capacity for conscious *Prohairesis* (Προαίρεσις).

The Fluidity of *Phronēsis* It is a necessary feature of the *Somatic logos* (Σωματικός Λόγος) that it creates synchronic snapshots—the *Demiourgos* and *Genetor* as distinct archetypes—to make this choice intelligible. However, *Physis* (Φύσις) expresses itself not as binary, static identities, but as fluid potentials.

The ideal Navigator is defined not by a fixed identity as a *Genetor* or *Demiourgos*, but by the cultivated capacity for **Phronēsis**—the automatic,

---

[19] See Section G.5 for the complete breakdown of the Popper Protocol and its implementation within the Navigator framework.

context-sensitive wisdom to discern which mode is demanded by the specific challenge within the *Archē*.²⁰ Through repeated *Energeia*, the *Logistikon* internalizes the *Logos* of situations, transforming conscious *Prohairesis* into fluent, intuitive mastery. The modes are not cages for the spirit, but tools for the Navigator.

8.6.3 Effective Action (*Eudaimonia* [εὐδαιμονία]) as the Output of a Coherent System

The ultimate output of this realigned state is effective action, which constitutes the naturalist interpretation of *Eudaimonia*—human flourishing.²¹ A system in *Energeia* is one where the *Logistikon* successfully orchestrates the *Psyche*: it strategically deploys the expressions of the *Hormē* (through *Orexis* and *Thymos*) toward long-term goals that are in alignment with the understood *Logos* of both the world and the self. This results in actions that are more adaptive, creations that are more functional, and decisions that yield better long-term outcomes. *Eudaimonia* is not a feeling of happiness, but the objective flourishing of a life well-navigated.²² It is the successful outcome produced by a *Psyche* in a state of *Dikaiosynē*.²³

8.6.4 The Continuous Engagement with the *Apeiron* (Ἄπειρον)

Finally, *Energeia* represents a new relationship with the *Apeiron*. No longer a source of terror or paralyzing confusion, the permanently unknowable is accepted as the permanent horizon for all action and exploration. The realigned *Nous* understands its purpose not as conquering the *Apeiron*, but as the continuous, active conversion of the **not-yet-known Archē** into the **known Archē**.

---

20. See Aristotle, *Nicomachean Ethics*, Book VI. *Phronēsis* is the ultimate navigational skill: the fluid switching between modes based on environmental feedback.
21. See Martha C. Nussbaum, *Creating Capabilities* (Cambridge, MA: Harvard University Press, 2011), 17–45. NPN defines *Eudaimonia* not as a feeling, but as the *capability* for effective action.
22. Cf. James Griffin, *Well-Being* (Oxford: Clarendon Press, 1986). Objective flourishing implies that *value* is a measure of functional success in the world, not subjective satisfaction.
23. In this state, effective action flows increasingly from *Phronēsis* (practical wisdom)—the cultivated capacity for discerning and executing the appropriate mode of engagement—rather than from conscious *Prohairesis* alone.

8.7 Conclusion: The Integrated Model

The theoretical architecture is now operational. We have mapped the *Psyche* and the *Archē*, but in this chapter, we have set the system in motion. We have defined the specific, algorithmic process by which a finite *Nous* self-corrects and navigates a dynamic reality.

The "Art of Navigation" is revealed not as a metaphor, but as a rigorous protocol. It is defined by the courage to endure the *Lysis* of our cherished *Doxa*, the discipline required to map the unknown territory encountered during *Aporia*, and the conscious exercise of *Prohairesis*. This choice requires discerning the moment for the stabilizing order of the *Demiourgos* mode versus the generative novelty of the *Genetor* mode. The culmination of this process is not a final answer, but the *Phronēsis* to live a life in *Eudaimonia*—a life of active, skillful alignment with the *Logos* and the flourishing it provides.

With this methodology established, we possess more than just a theory; we possess a diagnostic instrument. We can now look back at the history of thought not as a museum of dead ideas, but as a series of navigational attempts—some heroic, some disastrous.

This lens reveals the tragedy of the Great Divergence. The Pre-Platonics were the first explorers of this territory, mapping the *Logos* of the *Archē* with brilliant, if nascent, intuition. Their project was cut short by a pivotal moment: the Socratic induction of a widespread, foundational *Aporia*. Standing in this void, the subsequent tradition made a fundamental *Prohairesis*.

Plato's Synthesis was a choice for the *Genetor* mode on a transcendent scale.[24] Confronted by the flux of the sensory *Physis* and the stability of the mind's Somatic *logos*, he generated a new model: a realm of perfect Forms. This was a monumental act of philosophical creation, but one that

---

24. For the argument that Forms are a response to flux, see W. K. C. Guthrie, *A History of Greek Philosophy*, vol. 4 (Cambridge: Cambridge University Press, 1975), 50–63. NPN argues Plato made the "wrong" *Prohairesis* by fleeing the *Archē* rather than navigating it.

located the true *Logos* **outside** the physical *Archē*, inverting the actual ontological relationship and founding the Idealist tradition.

Aristotle's Structural Analysis was a choice for the *Demiourgos* mode of total immanence.[25] He sought to categorize and master the constituents of the world. Yet, his system represents the ultimate synchronic flattening of reality. He took the diachronic interplay of *Philia* and *Neikos* and projected it onto the static grid of the Four Causes, creating a magnificent map of Being that could not ultimately account for the Becoming of the *Nous* itself, forcing a quasi-idealist element back into his system.

The subsequent history of philosophy—the conflict between Idealism and Materialism—can be understood as the varying, and ultimately failed, attempts to resolve this core tension without the key of diachronic emergence.[26] Idealism correctly prizes the phenomena of mind but misplaces their origin. Materialism correctly identifies the fundamental ground but has lacked a compelling account of the navigator that stands upon it.

Neo-Pre-Platonic Naturalism resolves this by providing the missing diachronic foundation.[27] It grounds the *Nous* and its logic in the evolutionary history of the *Archē*.[28] The stage is now set to apply this lens with precision. In the chapters that follow, we will trace this journey in detail, from the recovery of the first map to the great divergence and its consequences, armed with the compass we have just forged (see Appendix G for the complete logical derivation of this operational protocol).

---

25. For the critique of Aristotle's substance ontology, see Alfred North Whitehead, *Process and Reality* (New York: The Free Press, 1978), 39–41. Aristotle froze the dynamic *Archē* into static categories, creating the "Synchronic Trap" NPN seeks to escape.

26. See Frederick Copleston, *A History of Philosophy*, 9 vols. (London: Burns, Oates & Washbourne, 1946–1975).

27. The notion of *Energeia* aligns with John Dewey's pragmatist epistemology. See *The Quest for Certainty* (New York: Minton, Balch & Company, 1929), 196–221. Truth is validated by the success of the action it directs.

28. This integrated model finds a parallel in Maurice Merleau-Ponty, *Phenomenology of Perception* (London: Routledge, 1962). The *Navigator* is an embodied consciousness, not a detached spectator.

Table 8.2 The Navigator's Path: from *Doxa* to *Eudaimonia*

NPN Term	Concept	Navigational Function
Eukairia (Εὐκαιρία)	Receptive Moment	The precondition for inquiry. A state where the *Thymos* is not defensive and the *Logistikon* is curious.
Elenchus (Ἔλεγχος)	Diagnostic Test	A rigorous, systematic interrogation of belief structures to expose contradictions within the current model.
Lysis (Λύσις)	Dissolution	The structural failure. The cognitive breaking down of foundational assumptions triggered by contradiction.
Exaiphnes (Ἐξαίφνης)	The Rupture	The sudden "Aha!" moment of profound disorientation; the realization that certainty was an illusion.
Katharsis (Κάθαρσις)	Purification	The conscious choice to release attachment to the falsified belief, clearing the *Psyche* for inquiry.
Aporia (Ἀπορία)	The Void	A state of conscious ignorance where old paths are closed, forcing the mind to confront the unknown without a map.
Dikē (Δίκη)	Re-anchoring	The acceptance of constraints. Realizing the *Archē* operates by impersonal laws regardless of one's wishes.
Prohairesis (Προαίρεσις)	Choice	The commitment to a mode of engagement— either Structural Analysis (*Demiourgos*) or Synthesis (*Genetor*).
Energeia (Ἐνέργεια)	Being-at-Work	The active implementation. Sustained expression of the *Nous* deploying the new model as a living hypothesis.
Phronēsis (Φρόνησις)	Practical Wisdom	Automated alignment. The state where the *Logistikon* governs action in alignment with the *Logos* without requiring the active intervention of the *Nous*.
Eudaimonia (εὐδαιμονία)	Flourishing	The objective result. A life well-navigated, where the *Psyche* operates in harmony (*Dikaiosynē*) with reality.

Chapter 9

The Platonic Synthesis: A Neo-Pre-Platonic Naturalist Critique

"Unless... philosophers become kings in their cities, or those who are now called kings and rulers genuinely and adequately philosophize... there will be no rest from ills for the cities." – Plato[1]

9.1 Introduction:
From Socratic *Kathairesis* to Platonic Synthesis

This chapter analyzes the philosophical shift in Plato's Middle Dialogues, arguing that his response to the Socratic *Aporia* (Ἀπορία)—while brilliant and systematic—represents a fundamental misstep in ontology. Through a critical examination of the *Meno*, *Phaedo*, *Symposium*, and *Republic*, we will contend that Plato correctly identified key phenomena of human consciousness but erroneously explained them through a top-down metaphysics of transcendence. The Neo-Pre-Platonic Naturalist (NPN) critique, grounded in the primacy of the physical *Archē* (Ἀρχή), reveals how Plato's system, designed as a solution to moral and epistemic chaos, inadvertently creates a blueprint for rationalized tyranny by inverting the actual, immanent relationship between mind and world.

---

1. Plato, *Republic*, 473c–d. This famous dictum encapsulates the central NPN critique: the fusion of unaccountable power with a detached, metaphysical *Logos* creates the structural conditions for tyranny. In *Plato: Complete Works*, ed. John M. Cooper (Indianapolis, IN: Hackett Publishing Company, 1997).

9.2 Confronting the Void

The early, or "Socratic," dialogues culminate in a state of *Aporia*—a productive void or impasse, masterfully induced by the *Elenchus* (Ἔλεγχος)—a method of cross-examination.² This process of *Kathairesis* (Καθαίρεσις)—a deconstruction or dismantling—successfully reduces unexamined beliefs (*Doxa*, Δόξα) to rubble but offers no positive doctrine. Plato inherits this void and seeks to fill it. The Middle Dialogues represent his ambitious **Synthesis**—the project of composition and reconstruction.

This chapter will argue that while Plato's diagnostic of the human condition—our capacity for abstraction, our internal conflicts, our drive for understanding—was profound, his prescribed solution inverted the actual, immanent relationship between mind and world. By projecting the mind's own operations onto a transcendent screen, he laid the groundwork for a system that, in its quest for perfect harmony, justifies a profound and troubling form of control.

9.2.1 The Structural Insight and the Metaphysical Error

Before deconstructing Plato's specific arguments, it is crucial to recognize what he got profoundly right at a *structural* level. Plato intuited that determinate, imperfect, and changing particulars could not be self-grounding. They required a contrasting background—a "not-this"—against which their identity could be defined. This is the insight that would later be formalized as the **General Zero Principle (GZP)** and its ontological corollary, the **Zero Principle (ZP)**, which states that a determinate system requires an indeterminate complement.

Plato's catastrophic error was not in seeing this need for a contrast-field, but in *characterizing it*. Where the Pre-Platonic tradition (exemplified by Anaximander's *Apeiron*) correctly pointed to an *indeterminate* ground, Plato populated that ground with perfect, eternal, and *hyper-determinate* entities: the Forms. He applied the logic of contrast correctly but inverted

---

2. For a prime example of the *Elenchus* inducing aporia, see the conclusion of Plato, *Euthyphro*, 15e. The dialogue ends not with a definition of piety, but with the necessary destruction of Euthyphro's false confidence.

its content, creating a transcendent realm where an immanent, indeterminate field was required. The rest of his system—the theory of recollection, the devaluation of the senses, the political blueprint of the *Republic*—flows from this initial, magnificent misstep. For the complete analysis of Plato's relation to the Zero Principle and his attempt to articulate what would become the General Zero Principle (GZP), see Appendix D.

9.3 The First Misstep: *Anamnesis* and the Misplaced Origin of Knowledge

The *Meno* presents a seemingly pragmatic question—is *Aretē* (Ἀρετή), excellence or virtue, teachable?—which quickly collapses into the familiar Socratic *Aporia*.[3] Faced with this crisis of unjustified belief, Plato introduces the theory of **Anamnesis** (ἀνάμνησις)—recollection. He posits that the *Psyche* (Ψυχή) is immortal and that what we call learning is actually the recollection of knowledge acquired before birth. The famous demonstration with the slave boy, who is guided to "discover" a geometric truth, is offered as evidence for this pre-natal knowledge.[4]

NPN demonstrates why this move is a critical misdiagnosis. Plato correctly identifies a real phenomenon: the human capacity for *a priori* reasoning and the possession of **innate cognitive structures**—what NPN terms *genetic knowledge*.[5] This includes the foundational capacity for logic, mathematics, and pattern recognition. However, he catastrophically misattributes their origin. This capacity is not evidence of a transcendent past life but is better explained as an evolutionarily conserved, deep-time adaptation.[6]

---

3. For the aporia regarding the teachability of virtue, see Plato, *Meno*, 80a.
4. For the theory of *Anamnesis* and the slave boy demonstration, see Plato, *Meno*, 80d–86c. NPN reads this not as proof of reincarnation, but as the first documentation of innate cognitive priors.
5. For the concept of evolved cognitive modules and innate structures, see Steven Pinker, *The Blank Slate* (New York: Viking, 2002), 33–45. What Plato calls "recollection" is actually the activation of evolved neural architecture.
6. For a modern naturalistic account of innate cognitive structures as evolved adaptations, see Daniel C. Dennett, *Darwin's Dangerous Idea* (New York: Simon & Schuster, 1995).

It emerges from the brain's long-term interaction with the *Logos* (Λόγος)—the impersonal, lawful, and discoverable structure inherent in the physical *Archē*. The slave boy is not remembering a Form; he is booting up an evolved cognitive module for spatial reasoning.

This misattribution has profound consequences. By locating the source of knowledge in a transcendent past, Plato severs the essential feedback loop between the emergent *Nous* (Νοῦς)—the conscious intellect—and the *Archē*. Learning is no longer a process of iterative model-building and testing through interaction with a lawful reality, but a turning *away* from that reality to consult a pre-loaded, internal database. This establishes the foundational epistemological gesture of his entire system: truth is not discovered through empirical engagement with the *Archē*, but recollected through a retreat from it. The first brick in Plato's Synthesis is laid by walling off the mind from the world that produced it.[7]

The theory of *Anamnesis* is thus more than an odd epistemological claim; it is the direct consequence of Plato's inverted ZP. If the contrast-field for knowledge is a realm of perfect, pre-existing truths (the Forms), then learning cannot be a constructive engagement with the *Archē* but must be a retreat into a transcendent memory. The slave boy in the *Meno* is not building a model of spatial relations through interaction with a lawful reality; he is "remembering" a truth from a determinate background that should have remained indeterminate.

9.4 The Metaphysical Flight: The *Phaedo* and the Fear of the *Apeiron*

If the *Meno* misplaces the origin of knowledge, the *Phaedo* provides the full metaphysical justification for turning away from the world. Set in Socrates's death cell, the dialogue argues systematically for the immortality of the *Psyche*, framing the body (*Soma*, σῶμα) as a prison that corrupts

---

7. The full exposition of the theory of anamnesis is in Plato, *Phaedo*, 72e–77a. For a comprehensive critique arguing the theory fails to account for experience, see Dominic Scott, *Recollection and Experience* (Cambridge: Cambridge University Press, 1995), 12–20.

CHAPTER 9. THE PLATONIC SYNTHESIS

and deceives.[8] The philosopher's life is presented as a preparation for death—the final separation of the *Psyche* from the body.

Plato sharpens a dualistic divide between the visible, changing, and composite world of *Becoming* and the intelligible, eternal, and simple world of *Being*. The *Psyche*, being simple and akin to the unchanging Forms (*Eidē*, εἴδη), is argued to be immortal through arguments from cyclical opposites, recollection, and affinity.[9]

NPN identifies this as Plato's fundamental ontological error: the **reification of abstraction**. Plato's Theory of Forms is a projection of the mind's own abstractive power—its ability to generate universal concepts like "Equality"—onto a transcendent screen.[10] The Form is not a supernatural entity but a conceptual model generated by consciousness to navigate a world of relative similarities and differences. It is a *tool* of the *Nous*, not its destination.[11]

Furthermore, Plato profoundly misreads the *Physis*. He interprets the dynamic, generative flux of reality not as the lawful manifestation of the *Logos*, but as an inferior realm of illusion and decay to be escaped. His metaphysics is thus a **flight** from the true, immanent nature of reality. He pathologizes the body and the senses, inventing a mind-body problem where there is only a biological process of integration between different cognitive systems—the immediate appetites, the mid-term social drives, and the long-term calculative faculty.

The *Phaedo*'s arguments are not neutral deductions but willful constructions designed to resolve the terror of mortality (for mortality itself belongs to the *Apeiron*; "what happens after death?" is an area of inquiry

---

8. For the body as a prison, see Plato, *Phaedo*, 62b–67b. This marks the decisive split from the "Somatic logos"—the body is no longer the ground of reason, but its enemy.

9. For the affinity argument linking the soul to the unchanging Forms, see Plato, *Phaedo*, 78b–84b.

10. For cognitive scientific accounts of abstraction as a biological process, see Stanislas Dehaene, *How We Learn* (New York: Viking, 2020), 87–112. The brain *creates* universals to compress data; it does not *discover* them in a separate realm.

11. This functionalist reading stands in opposition to the metaphysical conclusions drawn from the affinity argument, Plato, *Phaedo*, 78b–84b.

where no observations can be made, thus no conclusions can be drawn) and the *Apeiron*.[12] Plato correctly diagnoses the *Psyche*'s capacity for abstract thought but frames not just death, but embodied life itself, as a problem to be overcome. This represents a fundamental rejection of the Pre-Platonic project of understanding the *Archē* on its own terms.

9.5 The Motivational Engine: The *Symposium* and the *Eros* of Escape

The *Phaedo* provides the metaphysical justification for turning away from the world, but the *Symposium* provides the psychological engine. In this dialogue, Plato articulates his theory of **Eros** (Ἔρως)—not merely as sexual desire, but as the fundamental force of attraction that drives all human striving. The "Ladder of Love," described by Diotima, presents a process of sublimation where *Eros* for a beautiful body is gradually refined into *Eros* for beautiful souls, laws, knowledge, and finally, the Form of Beauty itself.[13]

NPN identifies this as a brilliant description of the **Philia** (Φιλία)—the drive for synthesis, connection, and understanding—as it operates in the human psyche.[14] It maps the progression of consciousness from concrete, immediate concerns to ever more abstract and stable models of reality.

However, Plato again inverts its direction. He frames this drive not as a mechanism for deeper engagement with the complex, embodied world, but as a linear ascent *away* from it. The highest form of love becomes the contemplation of a static, non-physical abstraction, divorced from the generative but messy strife (*Neikos*, Νεῖκος) of the physical realm.

---

12. For the Pre-Socratic concept of the *Apeiron* as the boundless ground, see Anaximander, KRS 101B. Plato treats the *Apeiron* as a threat to be conquered by the determinacy of Forms, rather than the necessary ground of existence.

13. For Diotima's Ladder of Love, see Plato, *Symposium*, 210a–212a. NPN reinterprets this ascent as the evolution of *Hormē* from basic homeostatic drives (*Orexis*) to complex social and intellectual strivings (*Nous*).

14. For the NPN concept of *Philia* as a cosmic force of attraction and synthesis, see the analysis of Empedocles in Chapter 2.

CHAPTER 9. THE PLATONIC SYNTHESIS

The *Symposium* thus completes the psychological groundwork for the *Republic*. It portrays the philosopher's highest calling not as a wise stewardship of the immanent world, but as a nostalgic yearning for a transcendent one. The generative potential of *Eros* is co-opted into a system of metaphysical escape, providing the emotional and motivational fuel for the Philosopher-King's flight from the cave. The love that should bind us to the world and to each other is redirected toward a ghostly paradigm, making the subsequent political control of the physical world seem not just necessary, but holy.

9.6 The Political Synthesis: The *Republic* as a System of Rationalized Control

The theoretical architecture of the escape is now complete. The *Phaedo* severed the mind from the body, and the *Symposium* redirected the heart toward the abstract. Yet, Plato does not leave the philosopher in the contemplative void; he demands a return. The *Republic* is the systematic attempt to impose the static perfection of the Forms onto the dynamic flux of the *Polis*. It is here that the metaphysical error transforms into a political tragedy: the effort to force the unruly *Hylē* of society to conform to a blueprint drawn in the sky.

9.6.1 From *Aporia* to the Philosopher-King

Having fled the productive *Aporia* of the Socratic method for a transcendent Synthesis, Plato faces a consequential problem: who is qualified to wield this complete, top-down system? The answer is the **Philosopher-King**. This figure is the logical and political embodiment of Plato's shift—the individual whose *Psyche* has apprehended the transcendent *Logos* and is therefore granted unaccountable authority to structure the state according to its dictates.[15]

---

15. For the definition and role of the Philosopher-King, see Plato, *Republic*, 473c–e. This represents the ultimate victory of the *Demiourgos* archetype: the Craftsman who imposes order without consultation.

9.6.2 The Blueprint for the Enlightened Tyrant

The core innovation of the *Republic* is its fusion of this transcendent knowledge of the *Logos* with absolute political power. The Philosopher-King, whose *Nous* has apprehended the Form of the Good, is granted unaccountable authority to structure the entire state.

NPN argues this structure is not merely vulnerable to tyranny; it is its logical outcome.[16] By providing a metaphysical justification for unaccountable rule—the "Noble Lie" to control the populace,[17] the censorship of poetry to shape the *Thymos* (θυμός) or spirited part,[18] the abolition of the family for the Guardian class[19]—Plato has designed a tyranny that believes its own propaganda.

It is a system where the **Demiourgos** (Δημιουργός)—the craftsman archetype of control—is elevated to absolute ruler. The Allegory of the Cave culminates not in a mission of universal liberation, but in a warning that the liberated philosopher will be killed by the masses if he tries to free them—a tacit instruction to manage the cave, not dismantle it.[20]

9.6.3 The Fatal Flaw:
The Ungrounded "Good" and the Unchained *Logistikon*

The entire system hinges on the ruler's knowledge of the "Form of the Good." However, this Form is a contentless abstraction with no necessary connection to the immanent well-being of conscious beings.[21] It can be filled with any content the ruler's *Logistikon* (λογιστικόν)—the calculating faculty—deems fit, from eugenics to totalitarian control. Plato's belief that

---

16. This structural critique of utopian, unaccountable rule is famously elaborated in Karl Popper, *The Open Society and Its Enemies*, Vol. 1 (Princeton, NJ: Princeton University Press, 1945), chap. 8. Popper argues that the attempt to arrest change inevitably leads to totalitarianism.

17. For the "Noble Lie," see Plato, *Republic*, 414b–415d. For analysis of this as a proto-totalitarian measure, see Popper, *The Open Society*, 138–45.

18. For the censorship of poetry to shape the *Thymos*, see Plato, *Republic*, 377a–392c.

19. For the abolition of the family for the Guardian class, see Plato, *Republic*, 457c–465c.

20. For the Allegory of the Cave and the philosopher's return, see Plato, *Republic*, 514a–521b.

21. For critiques of the Form of the Good as an empty abstraction, see Julia Annas, *An Introduction to Plato's Republic* (Oxford: Clarendon Press, 1981), 242–60.

the philosopher will be *morally compelled* to act for the benefit of all is the system's fatal act of faith. It is a wishful thinking that power, once fused with a *Logos* detached from the emergent dynamics of *Philia* and *Neikos* in the *Archē*, will somehow remain benevolent.

9.6.4 The Late-Dialogue Panic: Attempting to Cage the Monster

The arguments of *Republic* Books VIII–X function as a tacit admission of the system's inherent danger.[22] They represent a retrospective, and ultimately insufficient, effort to install safeguards against the tyrannical potential logically entailed by the model of the Philosopher-King. Having built a machine for perfect, enlightened control, Plato seems to realize he has handed a more sophisticated toolkit to the potential tyrant.

His subsequent efforts to "cage" this monster are tellingly weak:

- **The Misery of the Tyrant (Book IX):** A psychological ploy to argue that the unjust ruler lacks internal *Dikaiosynē* (Δικαιοσύνη)—the functional harmony of the *Psyche*—and thus cannot achieve *Eudaimonia* (εὐδαιμονία).[23] It is an appeal to self-interest, not a structural constraint.
- **The Myth of Er (Book X):** A post-mortem system of cosmic *Dikē* (Δίκη)—impersonal justice—a *deus ex machina* intended to enforce the behavior his political system fails to structurally guarantee.[24]

These appendices are a profound admission of failure. They reveal that the political structure *itself* contains no internal check on power. The only safeguards are appeals to the ruler's self-interest and fear of supernatural punishment—a stark contrast to the rigorous logical architecture of the state's control mechanisms over the populace. Having created a system for tyranny, Plato spends the final act pleading with it to be a Navigator and begetter of flourishing.

---

22. For scholarly analysis of these books as safeguards, see C.D.C. Reeve, *Philosopher-Kings* (Princeton, NJ: Princeton University Press, 1988), 235–60.
23. For the argument that the tyrannical man is the most miserable, see Plato, *Republic*, 576b–580c.
24. For the Myth of Er, see Plato, *Republic*, 614a–621d.

9.7 Conclusion: The *Republic* as a Monumental Warning

Neo-Pre-Platonic Naturalism concludes that the Platonic *Synthesis*, spanning the *Meno, Phaedo, Symposium,* and *Republic,* stands not as a utopian ideal, but as philosophy's most profound and systematic warning. Plato's genius lay in his unparalleled diagnostic—the tripartite map of the *Psyche,* the motivational power of *Eros,* the need to move from *Doxa* to *Epistēmē,* and the real phenomenon of innate cognitive structure. His catastrophic error was his prescription. Beginning with the misstep of *Anamnesis,* fortified by the flight from the *Apeiron* in the *Phaedo* and the otherworldly *Eros* of the *Symposium,* he fled the productive *Aporia* into a transcendent *Synthesis.*

This system, in seeking to eliminate the risk of ignorance and conflict of the masses, created a far more dangerous paradigm of unaccountable, self-justifying power. The Philosopher-King is not the solution to tyranny; he is its most perfect and terrifying embodiment: a *Nous* armed with a detached *logos,* driven by a deracinated *Eros,* and unmoored from the emergent truth of the *Archē*—a tyrant who rules the cave not by shattering its chains, but by perfecting its illusions.

The true philosophical task is not to follow this "Middle Plato" out of the cave, but to turn, with sober clarity, and learn to navigate the very world he sought to escape. **Remarkably, the first thinker to attempt this courageous turn was Plato himself.** The story of the *Republic* is not the end of his journey; it is the high-water mark of his error. In his final years, confronting the silence of the *Apeiron* and the stubborn reality of the *Polis,* the aging master would pick up the hammer of the *Elenchus* once more—this time, to dismantle his own castle in the sky. It is to this great unbuilding, and the emergence of the true Navigator, that we now turn.

Chapter 10

The Navigator and the Labyrinth: How Plato's Late Thought Forged the Way

"If then, Socrates, in many respects concerning many things— the gods and the generation of the universe—we prove unable to render an account at all points entirely consistent with itself and exact, you must not be surprised." – Plato[1]

10.1 Introduction: The Two Platos

The figure of Plato stands as a colossus over the Western intellectual tradition. For most, his name conjures a specific and powerful image: the idealist philosopher of eternal Forms, the architect of the Philosopher-King, the author of the *Republic*'s perfect, static city. This is the Plato who built a castle in the sky, inviting us to escape the shadowy cave of sensory deception for a realm of pure, transcendent truth. This portrait, however, is a profound and consequential misrepresentation. It captures the philosopher in middle age, brilliant and systematic, yet frozen in time. It ignores the deeper, more radical journey of his final years—a journey that constitutes his true, and largely unacknowledged, masterpiece.[2]

---

1. Plato, *Timaeus*, 29c–d. This quote serves as the charter for NPN's *Probabilistic Epistemology*: the acknowledgment that a finite *Nous* can only produce a "likely story" (*eikos mythos*) about the infinite *Archē*. In *Plato: Complete Works*, ed. John M. Cooper (Indianapolis, IN: Hackett Publishing Company, 1997).

2. The developmental interpretation of Plato, which sees a significant evolution in his thought from the middle to the late dialogues, is a central strand of modern scholarship. NPN aligns with this view, treating the late dialogues not as a decline, but as a critical maturation. See W. K. C. Guthrie, *A History of Greek Philosophy, Vol. 5: The Later Plato and the Academy* (Cambridge: Cambridge University Press, 1978), esp. 1–24, 415–20.

To be clear, this is not to posit two entirely separate Platos; the seeds of his late-stage critical methods can be found in earlier works, and the impulse for transcendent order never fully vanishes. Yet, the decisive shift in his final period—from a primary reliance on transcendent certainties to a focus on immanent structures and pragmatic models—is so pronounced that it effectively constitutes a second philosophical system built upon the ruins of the first.[3]

This chapter advances a decisive thesis: Plato's greatest and most enduring contribution emerges not from the confident system-building of his middle period, but from the critical, pragmatic, and immanent philosophy of his late dialogues. In a profound intellectual metamorphosis, the aging Plato systematically dismantled his own earlier doctrines, charting a course away from transcendent certainty and toward a philosophy of dynamic navigation of *Physis*. This course, we will argue, finds its natural and necessary destination in the First Principles of Neo-Pre-Platonic Naturalism (NPN).

The journey we will trace is one of deconstruction and reconstruction. It begins with a *Kathairesis* (καθαίρεσις)—a deliberate dismantling of his own foundations in the *Parmenides*. It moves through the epistemological *Aporia* (Ἀπορία)—a fertile void of the *Theaetetus*, creating the necessary starting place for a new foundation. From this *Aporia*, a new model emerges in the *Sophist* and *Philebus*, one that maps the *Logos* (Λόγος)—the lawful structure of the *Archē* and posits *Nous* (Νοῦς)—consciousness as the *Genetor* (Γενέτωρ)—the begetter of *Philia* (Φιλία)—attraction within it. Finally, we see the pragmatic reckoning of this view in the *Timaeus* and *Laws*, where the demand for *Demiourgos* (Δημιουργός)—the craftsman certainty is abandoned in favor of pragmatic models for a society with less *Demiourgos* and that is more *Genetor* in nature.

---

3. For a detailed scholarly analysis supporting a significant late-period shift away from transcendent psychology toward a developmental and immanent ethical framework, see Terence Irwin, *Plato's Ethics* (Oxford: Oxford University Press, 1995), esp. chaps. 24–25. NPN identifies this as the move from *Idealism* to *Naturalism*.

This is not the story of NPN as an external critique of Platonism. It is the story of NPN as the fulfillment of Platonism's deepest and most mature impulse. By providing the First Principles of the Primacy of the *Archē*, NPN grants Plato's late-stage insights the grounded, naturalistic foundation they sought but lacked, allowing the symphony he composed in his wisdom to finally be played. We have spent two millennia gazing at his blueprint for a castle in the sky; it is time we learned to use the maps he left for navigating the world.

10.2 The Great Unbuilding (*Lysis*)

The late dialogues can be read as Plato's protracted, often agonized confrontation with the flaws in his own foundational principle. Having built his system on an inverted application of the Zero Principle (a determinate Forms-based *Apeiron*), he now subjects that foundation to the very scrutiny it cannot withstand. The *Parmenides* and *Theaetetus* do not merely critique specific doctrines; they perform the *Lysis* (dissolution) required when a metaphysical structure violates the logic of contrast it implicitly depends upon.

This phase of Plato's final thought is dedicated to **Kathairesis**—the systematic destruction of the architectural assumptions built in the middle dialogues. Having diagnosed the failures of the transcendent ideal, Plato first needed to clear the intellectual ground. This great unbuilding occurs across two key dialogues:

1. **The Ontological Purge (*Parmenides*):** This dialogue targets the **metaphysics** of the middle period, dissolving the logical coherence of the Theory of Forms. By exposing the flaws in "participation," it forces the realization that a top-down, reality-by-correspondence model is logically untenable.
2. **The Epistemological Purge (*Theaetetus*):** Following the collapse of the Forms, this dialogue addresses **knowledge** itself, demonstrating the failure of perception, mere opinion, and justified true belief to provide absolute certainty. It leaves the philosopher stranded

in **Aporia**—the necessary void required before any new foundation can be built.

Together, these dialogues perform the necessary **Lysis**—the structural dissolution that frees the *Nous* from the burden of certainty. The entire edifice of the **Transcendent Architect** is taken down, forcing philosophy away from the sky and back toward the immanent ground of the *Archē*.[4]

10.2.1 The *Parmenides*: The Demolition of the Transcendent

The scene is a confrontation between philosophical generations. A confident young Socrates, brimming with the theory of Forms, faces the formidable old Parmenides, a philosopher of monolithic, unshakable rigor. Socrates presents his doctrine: that there exist transcendent, perfect Forms, separate from our world, in which particular objects merely "participate." Parmenides, with the patience of a master, proceeds not to ridicule, but to dissect.

Through a series of devastating arguments, most notably the "Third Man Argument," he exposes the fatal logical flaws in the theory of participation. If a Form exists to explain a set of similar things, what explains the similarity between the Form and the particulars? A second, higher Form is required, leading to an infinite regress that dissolves the very explanatory power the theory was meant to provide.[5]

This is not an attack by an outsider. It is Plato performing a *Kathairesis*—a deliberate, controlled demolition of his own foundational doctrine.[6] The

---

4. This reading views the *Parmenides* and *Theaetetus* as a continuous program of self-criticism—a deliberate process of dissolving earlier assumptions to clear the ground for the mature synthesis. See Myles Burnyeat, *The Theaetetus of Plato* (Indianapolis: Hackett Publishing Company, 1990), Introduction. NPN identifies this as the necessary *Kathairesis* of the Idealist paradigm.

5. For the "Third Man Argument" and the critique of the Theory of Forms, see Plato, *Parmenides*, 132a–133a. NPN reads this as Plato's admission that the Theory of Forms fails to satisfy the Zero Principle: a determinate Form cannot ground other determinations without regress.

6. The interpretation of the *Parmenides* as a serious, self-inflicted critique is a cornerstone of the developmental reading. See Gregory Vlastos, "The Third Man Argument in the Parmenides," *The Philosophical Review* 63, no. 3 (1954): 319–49.

CHAPTER 10. THE NAVIGATOR AND THE LABYRINTH   121

dialogue serves as a profound *Lysis* (λύσις)—dissolution for the reader, systematically cracking the edifice of transcendent metaphysics. The core realization is inescapable: a top-down, reality-by-correspondence model is logically untenable. It cannot serve as a coherent foundation for understanding the *Archē*.

The NPN Synthesis: The *Parmenides* represents the necessary *Lysis* stage, the dissolution that must precede any new construction. It clears the intellectual ground of the "Demiourgos mode"—the desire for a pre-existing, perfect blueprint to which reality must conform. By demonstrating the incoherence of a transcendent framework, it forces philosophy to seek a new foundation, not in a separate realm of ideal blueprints, but within the immanent structure of the *Archē* itself.[7]

10.2.2 The *Theaetetus*: The Void of Knowledge

The dialogue begins with the most fundamental of epistemological questions: "What is *Epistēmē* (Ἐπιστήμη)—justified, true knowledge?" Socrates and the promising young mathematician Theaetetus explore three definitions: that knowledge is perception, that it is true judgment, and finally, that it is true judgment with an account (*logos*).

Each proposal is subjected to rigorous *Elenchus* (Ἔλεγχος)—cross-examination. Perception is shown to be relative and unstable. True judgment is revealed to be possible by chance, lacking justification. The final, most sophisticated definition—justified true belief—is also dismantled, as the dialogue probes the elusive nature of an "account" that could transform belief into knowledge. The result is not a new definition, but an impasse. The dialogue concludes in *Aporia*.[8]

---

[7]. The consensus view is that the technical critique in the *Parmenides* is Plato's own attempt to demonstrate the logical incoherence of the Theory of Forms as formulated in the middle dialogues. See Richard Kraut, *The Cambridge Companion to Plato* (Cambridge: Cambridge University Press, 1992), 51–64.

[8]. For the conclusion of the *Theaetetus* in a state of aporia, see Plato, *Theaetetus*, 210a–d.

The realization here is as stark as it is fertile: there is no certain, unshakable foundation for knowledge to be found within the resources of experience and discursive reason alone.[9] The *Theaetetus* systematically closes every apparent exit, leaving the seeker in a state of productive perplexity. This is the epistemological void, the necessary clearing away of inadequate answers.

The NPN Claim on *Epistēmē*: The *Aporia* of the *Theaetetus* is resolved not by finding a new, fragile definition within the same paradigm, but by a fundamental shift in the concept of *Epistēmē* itself. NPN posits that because the *Nous* is an emergent phenomenon within the *Archē* that must operate through finite models, it can never achieve a state of perfect, binary correspondence. The presence of the *Apeiron* (Ἄπειρον) as the ground of all potential ensures an inherent gap between the model and the reality it represents. Therefore, *Epistēmē* is not a binary state of "justified true belief," but a high-confidence model that has proven its functional alignment with the *Logos* of the *Archē* through iterative, successful prediction and interaction. It is a spectrum of justification, where confidence is built empirically, not a status conferred by transcendent correspondence.[10]

The NPN Synthesis: This deliberate induction of *Aporia* is the crucial precondition for the NPN framework. It demonstrates that *Epistēmē* cannot be built upon the shifting sands of perception or ungrounded belief. This state of *Aporia* forces the acceptance of a first principle—the non-negotiable primacy of the *Archē*—and the redefinition of knowledge around a gradient of confidence. The dialogue's failure is, for NPN, the

---

9. The *Theaetetus*'s exhaustive investigation into knowledge is widely interpreted as Plato's demonstration of the failure of foundationalist epistemology. For a definitive modern commentary, see Burnyeat, *Theaetetus*, 1–10, 225–41. NPN interprets this failure as the proof that knowledge is a gradient, not a binary state.

10. The philosophical consensus is that the *Theaetetus* deliberately ends in *Aporia* to demonstrate the inherent failure of the "Justified True Belief" standard. See Burnyeat, *Theaetetus*, Introduction.

beginning of a more robust and naturalistic epistemology, where the measure of knowledge is its utility in navigating the world, not its fidelity to an unattainable ideal.[11]

10.3 The New Foundation (*Synthesis*)

Having systematically dismantled the architecture of his earlier thought, Plato's late work turns from critical *Kathairesis* to a constructive project. The Great Unbuilding (*Lysis*) cleared the ground of the transcendent. The New Foundation (*Synthesis*) explores the positive framework that emerges from the ruins—a framework that describes a dynamic *Archē* with *Logos* and outlines the ethics for navigating it. This is no longer a philosophy of escape to a higher realm, but one of masterful engagement with the *Archē* itself.

10.3.1 The *Sophist*: Mapping the Labyrinth of the *Logos*

The *Sophist* confronts a problem that would be insoluble within a rigid, binary framework of Being and Not-Being: how can we speak falsely? To define the Sophist—a purveyor of falsehoods—Plato must first rehabilitate the concept of "what is not." The Eleatic Stranger, the dialogue's protagonist, achieves this by arguing that "Not-Being" does not signify absolute non-existence, but rather **Difference** (*thateron*). "What is not" is simply "what is other than."[12]

This pivotal move allows for a revolutionary ontology. *Physis* is not a monolithic block of "Being" but a dynamic, structured web where *what is* is defined by its relations of Sameness and Difference with other things that are. The Stranger analyzes the interweaving (*symplokē*) of the "Greatest Kinds"—Being, Sameness, Difference, Motion, and Rest—to reveal the *Logos* as a complex system of relational properties.[13]

---

11. This reading is central to the developmental interpretation of Plato's thought. See Burnyeat, *Theaetetus*, Introduction. The NPN framework resolves this *aporia* by asserting that *Epistēmē* is **functional alignment along a confidence gradient**, grounding the system in the primacy of the *Archē*.
12. For the analysis of Not-Being as Difference, see Plato, *Sophist*, 257b–259b.
13. For the interweaving of the "Greatest Kinds," see Plato, *Sophist*, 254b–255e.

The NPN Synthesis: Healing the Parmenidean Wound: This is Plato explicitly addressing Parmenides's logical necessity of non-contradiction. He demonstrates that the "Many" (difference/relation) is not a contradiction of the "One" (Being), but its necessary condition. In doing so, he heals the catastrophic divide Parmenides opened when he mistook the static, *a priori* certainty of the *Somatic logos* for the structure of reality itself. By reintroducing Difference into the heart of Being, Plato effectively maps the **Cognitive Neikos**—the force of differentiation—that allows the *Nous* to navigate a complex, multifaceted *Archē*.[14]

10.3.2 Plato's Collection and Division: The NPN Synthesis

In the *Sophist*, Plato provides the very tools NPN identifies as fundamental to engaging with the *Archē*. He elevates the forces of *Philia* and *Neikos* from mere social concepts to impersonal methodological principles:

- **Collection (*Synagogē*):** The application of **Cognitive Philia**. This is the act of gathering disparate particulars into a unified kind or form (*mian idean*), allowing the Navigator to see the "One" within the "Many."
- **Division (*Diairesis*):** The application of **Cognitive Neikos**. This is the act of carving nature "at the joints," making the necessary distinctions that give a model its resolution and utility.[15]

The *Sophist* thus provides the metaphysical and methodological map for the **Genetor Nous** to begin its work of navigation and creation within *Physis*, not in opposition to it.

---

14. This reading, which views the late dialogues as moving toward a generative, relational ontology, is a major thesis in contemporary scholarship. See Kenneth M. Sayre, *Plato's Late Ontology: A Riddle Resolved* (Princeton: Princeton University Press, 1983).

15. For the methodological statement on Collection and Division, see Plato, *Sophist*, 253d–e. This method becomes a hallmark of the late dialogues, moving from the search for a single, transcendent Form to a complex, relational mapping of conceptual structures.

CHAPTER 10. THE NAVIGATOR AND THE LABYRINTH

Table 10.1 Plato's Collection and Division: The NPN Synthesis

Feature	Collection (Συναγωγή): The Force of Union	Division (Διαίρεσις): The Force of Separation
NPN Correlate	*Philia* (Φιλία): The Drive to Cohere	*Neikos* (Νεῖκος): The Drive to Differentiate
Goal/Action	To gather diverse particulars into one kind or form (*mian idean*).	To systematically cut a collected genus into mutually exclusive species (*kat' eidē*).
Direction	Upward (Moving from many specific instances to one broad genus).	Downward (Moving from one broad genus to specific subclasses).
Starting Point	Diverse instances or particulars that share a common feature.	The established broadest genus or kind.
Process	Identifying the necessary common feature shared by particulars (The force of Union).	Repeated dichotomous cutting "along its natural joints" (The force of Differentiation).
Result	A unified genus (the starting point for the next step, *Neikos*).	A genuine definition of the target concept, expressed by the successive steps.
Role in *Sophist*	Establishes the category of Expertise or Art (*techne*) as the starting point.	Traces the Sophist through a series of cuts (e.g., Acquisitive Art → Hunting → Hunting for gain).

10.3.3 The Biological Record as the Only Map of Deep Time

From the NPN perspective, the fixation of the late Academy—and subsequently Aristotle—on the classification of species was not a retreat into trivia; it was a strategic necessity mandated by the available evidence. Lacking the modern tools of geology or genomics to measure deep time directly, the diversity of living forms provided the *only* accessible dataset that preserved the accumulated effects of *Philia* and *Neikos* over aeons. A biological species is essentially a standing wave of *Philia* (inheritance) maintaining its structural coherence against the differentiating pressure of *Neikos* (variation and environmental strife). By applying Collection and Division to the "natural joints" of living things, Plato and Aristotle were not merely categorizing; they were reading the only historical record of the *Logos* available to them—attempting to reverse-engineer the dynamic interplay of cosmic forces from the static evidence of their results.[16]

10.3.4 The *Philebus*: The Ethics of the Navigator

If the *Sophist* provides the map of reality, the *Philebus* provides the ethical compass for the journey. The dialogue's central question—whether the good life consists in pleasure (*Hedonē*) or intellect (*Nous*)—receives a nuanced answer that rejects a simple binary: neither alone, but a measured mixture of both. The cause of this good mixture, however, is identified as *Nous*. It is the ordering intelligence that imposes measure and limit upon the boundless flux of experience to produce a harmonious and flourishing whole.[17]

This ethical conclusion is grounded in a profound fourfold ontological scheme:

1. **The Unlimited (*Apeiron*):** The boundless, indeterminate realm of more-and-less (e.g., the raw potential for pleasure and pain).
2. **Limit (*Peras*):** The principle of definite measure, ratio, and number (an expression of the *Logos*).
3. **The Mixture:** The harmonious combination of the Unlimited and Limit (e.g., a specific, measured, and therefore good, pleasure).
4. **The Cause (*Aitia*):** The intelligence, the *Nous*, that brings about the mixture.[18]

---

16. This connects the methodological shift in the *Sophist* directly to Aristotle's later biological project. See Aristotle, *Parts of Animals*, 642b5–644b20. For the intellectual context of the late Academy, see Guthrie, HGP Vol. 5, 35–39.

17. For the conclusion that the good is a mixture caused by *Nous*, see Plato, *Philebus*, 64a–67b.

18. For the fourfold ontological scheme, see Plato, *Philebus*, 23c–27c.

CHAPTER 10. THE NAVIGATOR AND THE LABYRINTH

The NPN Synthesis: The *Philebus* is the explicit ethics of the Navigator. The good life is the result of the Nous actively applying its aligned *logos* (as Limit and measure) to the chaotic potential of the *Archē* to generate new structures of a flourishing life. The *Nous* is no longer a fugitive from the world but its masterful navigator and orchestrator. This represents the full emergence of an individual operating with *Energeia* (Ἐνέργεια)—being-at-work, whose *Eudaimonia* is a direct result of their conscious, creative application of the *Logos* to the materials of existence.[19]

10.4 The Pragmatic Reckoning

The trajectory of Plato's late thought, having moved from demolition through reconstruction, culminates in a final, pragmatic reckoning. In his last works, Plato confronts the ultimate implications of his matured system: the necessary concession of absolute certainty and the challenge of building a just society for beings who are not perfected Navigators. This final part examines the institutionalization of the *Logos* and the acceptance of human epistemological limits as the final, crucial steps in Plato's journey toward a philosophy of immanent engagement.[20]

10.4.1 The *Timaeus*: The "Likely Story" and the Retreat of the *Demiourgos*

The *Timaeus* presents a grand cosmological myth: a divine craftsman, the *Demiourgos*, looks to the eternal Forms and imposes mathematical order and reason (*Nous*) upon a pre-existing chaotic receptacle to create the cosmos. This seems, on its surface, to be a regress to the transcendent metaphysics of the middle period—and it is this literal, theological reading that has dominated much of the dialogue's reception.

---

19. This interpretation, which grounds *Eudaimonia* in the purposeful, active application of reason to the constraints of reality, aligns with a functionalist view of ethics. See Martha C. Nussbaum, *The Therapy of Desire* (Princeton, NJ: Princeton University Press, 1994), 13–47.

20. This synthesis of Plato's late works—connecting the retreat from absolute certainty to the construction of a more pragmatic state governed by law—is central to the developmental reading of his political thought. See Julia Annas, *Virtue and Law in Plato and Beyond* (Oxford: Oxford University Press, 2017), 136–65.

However, to read the *Demiourgos* as a metaphysical fact is to ignore the dialogue's crucial and consistent epistemological caveat. Timaeus repeatedly emphasizes that his account is merely a "likely story" or "probable account" (*eikos mythos*). He justifies this by pointing to the changing nature of the physical world, which, as a copy, can only be an object of *Doxa* (Δόξα)—opinion and not *Epistēmē*.[21]

This is not a failure of argument but a philosophical acknowledgment of **the limits of human knowledge when confronting ultimate origins**. The *eikos mythos* signals that the generative ground of the cosmos—the relation between the *Demiourgos*, the Forms, and the receptacle (*khōra*)—cannot be captured in a certain, determinate account. It is a tacit admission of what NPN identifies as the *Apeiron*: the indeterminate complement that Plato's earlier system tried to fill with the hyper-determinate Forms.

Crucially, Plato introduces the **receptacle (*khōra*)** as a third ontological kind alongside Being (Forms) and Becoming (particulars).[22] The *khōra* is described as "invisible and characterless, all-receiving, partaking in some most puzzling way of the intelligible and very hard to apprehend."[23] This shift from a realm of every possible Form (a positive, determinate totality) to an "all-receiving," "characterless" matrix represents Plato's late-stage move toward **negative indeterminacy**. Having confronted the infinite regress entailed by a fully determinate grounding principle in the *Parmenides* (the "Third Man" problem), he now gestures toward a ground that resists positive characterization—a move that aligns his thought more closely with Anaximander's *Apeiron* than with his own earlier Theory of Forms.

The NPN Synthesis: The *Timaeus* represents Plato's pragmatic break from the demand for *Demiourgos*-like certainty (Mode A). Its methodological stance aligns with the NPN view of knowledge as a high-confidence

---

21. Plato, *Timaeus*, 29b–d. NPN reads the *eikos mythos* not as an excuse for poor rigor, but as the first formal definition of **probabilistic modeling**—the only form of knowledge possible for a finite knower.
22. Plato, *Timaeus*, 48e–52d.
23. Plato, *Timaeus*, 51a–b.

model rather than certain correspondence. Plato is effectively mapping the **Apeiron boundary** of human knowledge: the point where the complexity of the *Archē* exceeds the capacity of the *Nous* to fully map it. While Plato frames this as a concession to the mutable nature of the cosmos, NPN asserts this boundary as a fundamental First Principle (FP5: Impotence Before the *Apeiron*). The *Demiourgos* thus retreats into myth, leaving the *Genetor Nous* to work with "likely stories"—probabilistic, functional models—while navigating the not-yet-known *Archē*.[24]

10.4.2 The *Laws*:
The Second-Best State and the Framework for Order

The *Laws*, Plato's final and longest dialogue, represents the ultimate pragmatic application of his late-stage thought. The project is the creation of a concrete legal code for a new city, Magnesia. Strikingly, the dialogue explicitly abandons the ideal of the Philosopher-King from the *Republic*. The Athenian Stranger, the dialogue's lead figure, admits that such perfectly wise rulers are a rarity, and to base a state on such a premise is unrealistic. Instead, he proposes the "second-best" state, governed not by the unfettered *Nous* of a king, but by the impersonal, codified rule of law.[25]

This shift is monumental. The law (*nomos*) is the institutionalization of the *Logos*. It is the accumulated, written wisdom of the community, designed to guide citizens toward virtuous behavior and harmonious (*Philia*-promoting) relationships in the consistent absence of perfect, Navigator insight. The law acts as an external scaffold, a pre-packaged *Logos*, intended to cultivate the internal *logos* of the citizenry. It is a system designed to generate right action, distributing a semblance of Navigator judgment across the population.

---

24. For scholarly analysis of the *eikos mythos* as a profound concession on the nature of physical explanation, see G. E. R. Lloyd, *Methods and Problems in Greek Science* (Cambridge: Cambridge University Press, 1991), 314–21.

25. Plato, *Laws*, 875a–d. Here, the Athenian Stranger explicitly argues that since a truly wise ruler is a rarity, the "second-best" course is to be ruled by the impersonal and fixed "commands of law," directly prefiguring the NPN concept of a *Demiourgos* framework that provides stability for a non-ideal populace.

Crucially, Plato recognizes that this legal scaffold cannot run on autopilot. In Book XII, he introduces the **Nocturnal Council** (*Nykterinos Syllogos*)—a select body of elders and auditors who meet before dawn.[26] Their specific mandate is to study the philosophical principles behind the laws—to know the 'Unity of Virtue' and the nature of the cosmos.[27] They are the city's living tether to the *Logos*. While the citizens live within the stabilizing mythos of the law, the Council acts as the hidden **Navigator core**, ensuring that the *Demiourgos* structure remains aligned with the dynamic reality of *Physis*. Without them, the state is a body without a soul, destined to drift into maladaption.[28]

The NPN Synthesis: The *Laws* is the societal recognition of the *Demiourgos/Genetor* distinction. It is an attempt to build a system that compensates for the scarcity of individuals operating in full *Energeia* as Navigators. By codifying the *Logos* into law, Plato seeks to create a structure that systematically reinforces *Dikaiosynē* (Δικαιοσύνη)—functional harmony and guides individuals toward Navigator-like alignment with the *Logos* of the *Archē*. It is the ultimate pragmatic outcome of his late philosophy: since we cannot all be Navigators, we must build a kingdom of laws that teaches us how to think and act like one.[29]

10.5 Conclusion:
The Unfinished Symphony and Its Conductor

The journey through Plato's late dialogues reveals not a scattered collection of works, but a coherent and radical intellectual project. It is the story of a philosopher who had the courage to dismantle his own most

---

26. Plato, *Laws*, 961a, 969b. The Council meets "daily from dawn until the sun rises."
27. Plato, *Laws*, 964a–965a. Plato insists the guardians must see the "one in the many" regarding virtue.
28. This concept finds a precise functional correlate in NPN's reading of the **Nocturnal Council**. It identifies this not as a mere political preference, but as a systemic necessity: the "Second-Best State" requires a specialized, philosophically aligned core (the Keepers of the Functional Mythos) to prevent the inevitable entropic drift of the legal scaffold.
29. The *Laws* explicitly proposes the **"second-best" state** ruled by law due to the rarity of a perfectly wise ruler (*Laws*, 875a–d). This functional view is supported by ethical scholarship on Plato's late political thought. See, for example, Annas, *Virtue and Law*, 136–65.

CHAPTER 10. THE NAVIGATOR AND THE LABYRINTH

famous doctrines and, in their place, lay the groundwork for a philosophy of dynamic, immanent engagement. The *Parmenides* and *Theaetetus* performed the necessary *Lysis*, clearing away the logically untenable structures of transcendent Forms and certain knowledge. From this *Aporia*, the *Sophist* and *Philebus* began the *Synthesis*, mapping the relational *Logos* of the *Archē* and defining the ethics of the Navigator *Nous* that navigates it. Finally, the *Timaeus* and *Laws* presented the pragmatic reckoning, conceding the limits of human knowledge and institutionalizing the *Logos* as law to guide a fallible humanity.

This dramatic evolution was almost certainly not an isolated endeavor. The presence of the young Aristotle in the Academy during this final period represents one of the most fruitful collisions of intellect in history. The relentless logical scrutiny of the *Parmenides*, the methodological rigor of the *Sophist*, and the turn toward a philosophy grounded in the nature of the physical world and political reality evident in the *Philebus* and *Laws* all read as a profound engagement with the systematic and immanent mindset of his brilliant student. The late dialogues, therefore, can be seen as Plato's ultimate philosophical testament, forged in the crucible of a dialogue with the thinker who would both inherit and transform his legacy.[30]

The tragic error of the Western tradition has been to privilege the blueprint of the middle-aged system-builder over the maps of the mature Navigator. For more than two millennia, we have been trying to inhabit Plato's castle in the sky, a pursuit that has led to endless philosophical dead ends and a deep alienation from the *Archē* itself.

---

30. While direct evidence is scarce, the chronological overlap of Aristotle's residency at the Academy with the composition of Plato's late works makes a relationship of mutual influence highly plausible. See Guthrie, *HGP Vol. 5*, 35–39.

Table 10.2 The Two Platos: The Shift from Architect to Navigator

Feature	Middle Period (The Architect)	Late Period (The Navigator)
Metaphysics	**Transcendent Idealism:** Reality is a shadow of perfect, static Forms located in a separate realm.	**Immanent Naturalism:** Reality is a dynamic *Physis* structured by an internal, relational *Logos*.
Epistemology	**Binary Certainty:** Knowledge is absolute, recovered via recollection (*Anamnesis*) of the Forms.	**Probabilistic Models:** Knowledge is a "Likely Story" (*Eikos Mythos*), a high-confidence map of the *Archē*.
Political Ideal	**The *Republic*:** Utopian rule by a Philosopher-King with unconstrained insight.	**The *Laws*:** Pragmatic rule by impersonal Law (The "Second-Best State") to guide fallible citizens.
Methodology	**System Construction:** Building a static "Castle in the Sky" (*Demiourgos* Mode).	**Critical Navigation:** Dismantling errors (*Lysis*) and mapping the territory (*Genetor* Mode).

"And yet, Socrates, if a man... does not allow that there are forms of things and does not define a form of every individual thing, he will have nothing on which his mind can rest, since he does not allow that at all times the form of each thing is the same, and in this way he will utterly destroy the power of discourse."
— Plato[31]

---

31. Plato, *Parmenides* 135b–c. Here Plato is breaking the momentum of his own critique to deliver a frantic warning. After spending the dialogue dismantling the immature version of the Theory of Forms, he pivots to reveal the terrifying alternative: **Nominalism**. He asserts that despite the logical difficulties of establishing the Forms, abandoning these structural invariances leaves the mind with "no place to rest." Without a stable, objective reference point—a fixed *Arche* or Principle—language itself (*logos*) loses its traction on reality, and the possibility of philosophy vanishes. He is not defending a dogma; he is identifying the structural precondition for intelligibility.

CHAPTER 10. THE NAVIGATOR AND THE LABYRINTH

Neo-Pre-Platonic Naturalism is not an attack on Platonism, nor does it diminish Plato's stature. It completes and elevates him, revealing the visionary who began to synthesize the core insights of the Pre-Platonics—the *Apeiron* of Anaximander, the *Logos* of Heraclitus, the dynamic forces of *Philia* and *Neikos* from Empedocles—into a coherent, immanent framework.

The late dialogues composed a profound symphony, but one that lacked a grounded conductor. By asserting the First Principle: Primacy of the *Archē*, NPN provides that conductor. It grants the mature Plato's thought the naturalistic foundation it sought, allowing the symphony he composed in his wisdom to finally be played. It is time to take Plato's journey from constructor of prisons to pioneer of reality more seriously, and to right this historic wrong.[32]

---

32. The developmental reading of Plato, which credits his late thought with a shift toward an immanent and systematic philosophy, is supported by Guthrie, *HGP Vol. 5*, who views this period as the convergence of earlier rational inquiries.

Chapter 11

The Aristotelian Impasse: Synchronic Mastery and the Unmoved Mover

"The actuality of thought is life." — Aristotle[1]

11.1 Introduction: The Completion of the Immanent Turn

The preceding chapter argued that Plato's late work constitutes an *immanent turn*—a deliberate dismantling of transcendent Forms and a tentative move toward a philosophy grounded in the dynamic, processual *Physis* (Φύσις). This chapter posits that Aristotle's project is the **systematic implementation—and ultimate synchronic flattening—of this turn.**[2] Where the late Plato began to chart the diachronic territory of a generative reality, Aristotle provided its definitive, yet static, map. He perfected the lexicon of immanence, but in doing so, he systematically erased the dimension of time and the primacy of Becoming that his teacher had begun to explore.

His core achievement is the hylomorphic synthesis, which seeks to account for all of reality—from inert matter to abstract thought—within a single, coherent, and synchronic framework. Yet, this very strength reveals the fundamental limitation of a purely synchronic worldview. From the

---

1. Aristotle, *Metaphysics*, 1072b26. NPN adopts this identification of *Energeia* (Actuality) with Life, but strips it of the theological necessity of an Unmoved Mover. In *The Complete Works of Aristotle: The Revised Oxford Translation*, ed. Jonathan Barnes (Princeton, NJ: Princeton University Press, 1984).

2. W. K. C. Guthrie, *A History of Greek Philosophy, Vol. 6: Aristotle: An Encounter* (Cambridge: Cambridge University Press, 1981), 25–27. Guthrie argues that Aristotle's philosophy is fundamentally an attempt to systematize and naturalize the insights of the later Plato within a fully immanent framework.

vantage of Neo-Pre-Platonic Naturalism, Aristotle's system represents the most rigorous possible attempt to construct a complete description of *Physis* as a static arrangement. However, this synchronic commitment inherently limits its ability to account for the diachronic, generative processes that constitute reality as a dynamic, historical Becoming. This limitation, which first appears as a crack in his metaphysics of mind, ultimately fractures his entire political project, revealing a system that perfects a static human essence but cannot cultivate a dynamic Navigator.

11.2 The Synchronic Summit: Mapping the Static *Archē*

With Aristotle, the project of Greek rationalism achieves its most rigorous and systematic expression. He does not merely observe the **Archē** (Ἀρχή); he catalogues, categorizes, and defines it with a precision that would dominate Western thought for two millennia. However, this clarity comes at a price. By prioritizing substance—**Ousia** (Οὐσία)—and fixed form, Aristotle constructs the ultimate map of a **Synchronic** reality—a world understood as a collection of static things rather than a flow of dynamic processes. This section explores the architecture of that magnificent, static map and the specific point where it fails to capture the living **Nous** (Νοῦς).[3]

11.2.1 The Hylomorphic Labyrinth: A Map of a Static World

Aristotle's system is a labyrinth of interconnected concepts, all grounded in the primacy of physical substance and the relationship between potentiality—**Dynamis** (Δύναμις)—and actuality—**Energeia** (Ἐνέργεια). To anchor this metaphysics, Aristotle famously co-opted the common Greek word for "wood" or "timber"—**Hylē** (ὕλη)—repurposing it as a technical term for "prime matter" or the underlying substratum that receives form.[4]

---

3. This interpretation of Aristotelian substance ontology as fundamentally static, misrepresenting a universe of dynamic processes as a collection of static subjects, is the cornerstone of process philosophy. See Alfred North Whitehead, *Process and Reality: An Essay in Cosmology* (New York: The Free Press, 1978), 39–41, 50–51.

4. For Aristotle's introduction of *hylē* as a technical philosophical term for matter, see Friedrich Solmsen, "Aristotle's Word for Matter," *Didascaliae* (1961): 395–408. NPN restores the "timber" metaphor to its active, growing roots.

CHAPTER 11. THE ARISTOTELIAN IMPASSE 137

In NPN terms, this distinction between Potentiality (*Hylē*) and Actuality (*Morphe*) is the first rigorous formalization of the **General Zero Principle (GZP)**. Aristotle correctly recognizes that determination cannot exist in a vacuum; it requires an indeterminate background to provide the necessary contrast. *Hylē* acts as the localized *Apeiron*—the passive indeterminacy that allows the active *Logos* to generate a specific "this" (*tode ti*) from the general potential. By lexicalizing the "timber" of the cosmos, Aristotle shifted the focus from the Pre-Socratic *Archē* (the living, active source) to a passive, synchronic material awaiting the imposition of Form. For the complete analysis of Aristotle and GZP, see Appendix E.

His definition of the ***Psyche*** (Ψυχή) as the "first ***entelecheia*** (ἐντελέχεια)— the state of having one's full reality or perfection—of a natural body having life in potentiality" is the master-key.[5] The *Psyche* is not a separate thing but the form and function of the living body. He replaces Plato's tripartite *Psyche* with a hierarchy of integrated powers (*Dynameis*): the nutritive—***Threptikon*** (Θρεπτικόν), the perceptive—***Aisthētikon*** (Αἰσθητικόν), and the intellectual—***Dianoētikon*** (Διανοητικόν).[6] This is a systematic naturalization of psychology, translating a political-psychic structure into a biological-functional one.

NPN Analysis: Aristotle's hierarchy is a brilliant synchronic snapshot of the Navigator's evolved architecture. The NPN framework interprets these *Dynameis* not as static powers, but as layered, diachronically-selected competencies. What Aristotle describes as a static hierarchy, NPN understands as the integrated toolset of the emergent navigator. By coining *Hylē*, Aristotle provided the "lumber" for a world of static objects, but NPN restores the "living timber" by viewing matter as a process bounded by the *Apeiron*.

---

5. Aristotle, *De Anima*, 412a27. This definition naturalizes the soul, removing it from the category of "ghost" and placing it into the category of "biological function."
6. For the hierarchy of the soul's faculties, see Aristotle, *De Anima*, 414a29–415a13.

While Aristotle conceived of *entelecheia* as the static perfection of a fixed essence, its functional equivalent in the NPN diachronic reality is **Eudaimonia** (εὐδαιμονία). *Eudaimonia* is the objective flourishing of a life well-navigated—the successful outcome produced by a *Psyche* in sustained *Energeia*, which serves as the metric for a life lived in alignment with the cosmic **Logos** (Λόγος).[7]

11.2.2 The Flattening of Reality: *Philia, Neikos*, and the Four Causes

From the NPN vantage, Aristotle's system represents the most comprehensive attempt in history to 'flatten' the diachronic nature of reality into a synchronic interpretation.[8] His method is one of heroic simplification: he takes the four-dimensional, turbulent process of *Physis*—the very interplay of **Philia** (Φιλία) and **Neikos** (Νεῖκος) that the Pre-Socratics identified—and projects it onto a static, logical grid where it can be mapped, categorized, and understood.

His entire project is an effort to institutionalize these dynamic, pre-conceptual forces into a static framework. His method of biological classification is a sustained act of cognitive *Neikos*—the force of separation, drawing precise boundaries between species. Yet, the goal of this division is to understand the shared, unifying *logos* of a species—the work of *Philia*.

This interplay is crystallized in his doctrine of the Four Causes, which can be understood as a synchronic snapshot of a diachronic process:

- **The Final Cause** freezes the future goal—the *telos* (τέλος) that pulls a process into being—into a timeless, pre-existing attractor. It is *Philia* conceived as a static destination.

---

7. For the interpretation of *Eudaimonia* as the objective highest human good achieved through excellent rational activity, see Martha C. Nussbaum, *The Fragility of Goodness* (Cambridge: Cambridge University Press, 2001), 317–42. NPN adopts this objectivist stance: flourishing is a fact, not a feeling.

8. Whitehead, *Process and Reality*, 39–41, famously critiqued the Aristotelian substance ontology for its "static" quality. NPN identifies this not as an error, but as a methodological trade-off: Aristotle bought precision at the cost of time.

CHAPTER 11. THE ARISTOTELIAN IMPASSE

- **The Formal Cause** reifies an emergent pattern—the unified structure that arises from interaction—into an eternal essence. It is the product of *Philia*'s unifying power, treated as a prior blueprint.
- **The Efficient Cause** reduces the historical chain of events and the violent friction of change to a single, prime mover. It captures the moment of *Neikos*—the disruptive force that breaks the old unity—but isolates it from the continuous flow of time.
- **The Material Cause** treats the dynamic, transformative *Hylē*—the realm of potentiality and strife—as a passive, inert substrate. It is *Neikos* neutralized into mere stuff.

This is not an error; it is a methodological masterstroke that allows for the creation of a complete, coherent map of 'what is.' He is the supreme cartographer of being. But his map, for all its genius, systematically erases the dimension of time and the generative processes of Becoming. He describes the stable structures created by *Philia* and *Neikos*, but his synchronic lens cannot capture the reality of the territory-making process itself.[9]

> *"Knowledge is the object of our inquiry, and men do not think they know a thing till they have grasped the 'why' of it (which is to grasp its primary cause)."* — Aristotle[10]

---

9. For the doctrine of the Four Causes, see Aristotle, *Physics*, 194b16–195b30. The view that this framework constitutes a "static flattening" of reality is the central critique of philosophers of becoming. See Whitehead, *Process and Reality*, 39–41, 50–51.

10. Aristotle, *Physics*, 194b17–20. This definition highlights the central tension between the two systems. NPN agrees that true knowledge (*Epistēmē*) requires grasping the "why" (causal necessity) rather than merely the "that" (observation). However, NPN diverges on the nature of that cause: where Aristotle locates the "why" in static, eternal structures (Forms/Telos), NPN locates it in the dynamic, diachronic pressure of *Hormē* and the homeostatic regulation of the system over time. For NPN, the "primary cause" is not a fixed blueprint, but the active process of self-maintenance.

Table 11.1 The Synchronic Flattening: Mapping Causes to Forces

Aristotelian Cause	NPN Correlate	The Process of Flattening
Final Cause	*Philia* (Static)	Freezes the future goal (*Telos*) into a timeless, pre-existing attractor.
Formal Cause	*Philia* (Structural)	Reifies emergent patterns of unification into a fixed, eternal essence or blueprint.
Efficient Cause	*Neikos* (Punctual)	Isolates the disruptive, historical friction of change into a single, momentary "mover."
Material Cause	*Neikos* (Neutralized)	Reduces the dynamic, generative strife of *Hylē* into a passive, inert substrate.

11.2.3 The Crack in the Foundation: The *Aporia* of the *Nous*

This elegant, flattened system fractures when confronted with the faculty responsible for the act of flattening itself: the human intellect. To account for the mind's ability to "become all things," Aristotle introduces his famous distinction. The passive intellect (*Nous Pathētikos*) fits within the hylomorphic scheme.[11] But the active intellect (*Nous Poietikos*) is described in terms that shatter it: "separable, unmixed, and impassible," being in its essence pure *Energeia*, "and this alone is immortal and eternal."[12]

Here, the Aristotelian drive toward total immanence meets its limit. Having so brilliantly systematized *Philia* and *Neikos* everywhere else, he encounters a faculty—the generator of form and the ultimate unifier—that

---

11. Aristotle, *De Anima*, 430a10–25.
12. Aristotle, *De Anima*, 430a17–18. This "immortal spark" is the ghost in the machine that resists total naturalization.

CHAPTER 11. THE ARISTOTELIAN IMPASSE   141

resists this very systematization. The *Nous Poietikos* is the **Unmoved Mover within the microcosm**, a divine, transcendent spark necessary to actualize thought. This concept represents a fundamental violation of the **General Zero Principle**. By defining the Active Intellect (and the Unmoved Mover it mirrors) as "Pure Actuality"—total determination with zero potentiality—Aristotle attempts to posit a Figure without a Background. He constructs a system with an "inside" but no "outside," asserting a determinate identity that relies on no contrasting indeterminacy. This violation forces him to extract the *Nous* from the fabric of the *Archē*, rendering it a "ghost" that breaks the hylomorphic unity he labored to build. It is the ghost of Platonism within the Aristotelian machine—the one diachronic process that refuses to be flattened into a synchronic substance.[13]

11.3 The NPN Resolution: From Static Substance to Diachronic Becoming

The contradiction at the heart of Aristotle's psychology—the struggle to explain how a passive, biological organ can possess an immortal, active intellect—cannot be resolved within a Synchronic framework. To save the phenomenon of the mind, we must abandon the static metaphysics of substance and embrace the **Diachronic Primacy** of NPN. We do not need to posit a divine spark inserted from without; we need only to recognize the emergent properties of a complex system operating across time.

11.3.1 The *Nous* as Emergent *Energeia*: Reframing the Active Intellect

Aristotle's *aporia* is the inevitable result of seeking a synchronic cause for a diachronic faculty. The NPN framework dissolves this contradiction through a decisive re-orientation: the *Nous* is not a "thing" to be defined,

---

13. For a classic analysis of the active intellect as a problematic exception to Aristotle's hylomorphism, see Franz Brentano, *The Psychology of Aristotle* (Berkeley: University of California Press, 1977). Brentano highlights how the "separable" intellect breaks the unity of the soul-body composite.

but a process—the emergent activity of a navigator. The so-called "active" intellect is not a separate activator but the *Energeia* of the modeling process itself.

This reframing resolves the Aristotelian dilemma in three moves:

1. **The *Nous Poietikos* is Emergent *Energeia*:** It is the generative capacity of the entire cognitive system operating at its highest level, not a separate substance.
2. **The Unmoved Mover is the *Logos*:** Aristotle's first principle is best understood not as a conscious entity, but as the *Logos* itself—the lawful structure of the *Archē* that acts as the ultimate attractor state, conditioning all process and Becoming.
3. **The Four Causes as the Navigator's Schema:** The Four Causes are reframed not as a static list but as the schema of the Navigator's engagement. The *Archē* provides the constraints (Material/Efficient Causes), and the Navigator provides the direction (Formal/Final Causes). This is the logic of the *Genetor*.

11.3.2 The NPN Analysis: *Eudaimonia* as the Navigator's True North

Aristotle's identification of *Eudaimonia* as the human *telos* is the heart of the NPN project. He correctly identified the target state: a life of flourishing achieved through the excellent activity (*Energeia*) of a rational being.[14] This is the core First Principle of a successful navigation.

Where NPN diverges is not in the destination, but in the model of the traveler and the map. Aristotle's system provides a static, synchronic portrait of the Navigator in a state of achieved harmony. It is a masterful description of the **harbor**—a place where the internal *logos* of the well-ordered *Psyche*, the external *Logos* of the just *polis*, and the cosmic *Logos* of the Unmoved Mover are in perfect, resonant alignment. His virtue ethics is the blueprint for maintaining this harmonious state.

---

14. For *Eudaimonia* as the highest human good and activity in accordance with virtue, see Aristotle, *Nicomachean Ethics*, 1098a7–18.

The NPN framework, by contrast, is the art of navigation for a journey that is inherently diachronic and often through uncharted waters. It accounts for the reality that the Navigator is not a static essence but an evolving, self-correcting system. The *logos* of the *Nous* (the internal map and goals) and the *Logos* of the *Archē* (the external territory and its laws) can and do conflict. Aristotle's ethics, lacking this distinction and the diachronic forces of *Philia* and *Neikos*, possesses no robust mechanism for resolving such fundamental conflicts. It describes the state of *being* aligned, but not the often-traumatic process of *Becoming* aligned through the iterative testing of models against the hard rock of reality.

11.4 The Political Consequence: The Cost of a Flattened Reality

Metaphysics is never politically neutral. If one models the *Archē* as a static hierarchy of fixed forms, one will inevitably model the *Polis* as a rigid hierarchy of fixed classes. Aristotle's political theory is the logical downstream consequence of his Synchronic flattening. Because he views the "nature" of things as a fixed destination to be reached, he views the State not as a vehicle for exploration, but as a mechanism for molding human material into its "proper" shape.

11.4.1 The Aristotelian Polis: The State as Perfecter of a Fixed Nature

Aristotle's political philosophy is the macro-scale application of his synchronic framework. Beginning with the axiom that "man is a political animal," he deduces that the *polis* is the natural and necessary culmination of human association, the entity that exists for the sake of the "good life" (*to eu zēn*), which is *Eudaimonia*.[15] His system classifies constitutions based on the number of rulers and whether they govern for the common good or

---

15. For man as a political animal and the *polis* as existing for the sake of the good life, see Aristotle, *Politics*, 1253a1–39. See also David Roochnik, "What Is Natural about Aristotle's Polis?," *Polis* 26, no. 2 (2009): 217–40.

their own interest, with his pragmatic ideal being the mixed constitution of "Polity," ruled by a stable middle class under the rule of law.[16]

The core of his vision is that the state is the primary agent for actualizing a fixed human potential. It provides the education, structure, and stable environment required to mold individuals into their essential, virtuous form. The *polis* is the craftsman, and the citizen is the material brought to its perfected state. This is a top-down, structural solution to human flourishing, a magnificent and coherent blueprint for a static world.

11.4.2 The Synchronic Blind Spot: The "Natural Slave" as a Case Study in Flattening

The profound ethical and political consequences of this 'flattening' are nowhere more starkly revealed than in Aristotle's doctrine of the 'natural slave.' This is not merely a reflection of his cultural prejudice, but a direct and chilling application of his synchronic method to human beings.

A strictly literal reading of the *Politics* confirms his defense of the chattel slavery of his era.[17] However, the NPN lens reveals a deeper, more systemic error. His method, which reads ontological status directly from observable function, provides no conceptual space for a dormant potential that could fundamentally alter an observed state. **The synchronic snapshot is the reality.**[18]

From this constrained vantage point, Aristotle's "natural slave"—defined as one who "participates in reason enough to apprehend it but not to possess it"—is the ultimate result of his epistemic flattening.[19] He observes a human being whose rational capacity, at the time of observation, is only applied to the execution of another's *logos*. His taxonomic imperative, the

---

16. For the classification of constitutions and the mixed regime, see Aristotle, *Politics*, 1279a22–1279b10, 1293b22–1294b41.
17. For the doctrine of the natural slave, see Aristotle, *Politics*, 1254b16–1255a3.
18. The doctrine's conceptual origin as a logical necessity within Aristotle's system is analyzed in Malcolm Schofield, "Ideology and Philosophy in Aristotle's Theory of Slavery," in *Aristoteles' "Politik"* (Göttingen: Vandenhoeck & Ruprecht, 1990), 1–27. Schofield demonstrates that the "Natural Slave" is the inevitable result of applying functional teleology to human beings.
19. Aristotle, *Politics*, 1254b22–23.

very *Neikos* that so brilliantly categorized the natural world, now categorizes a person. The conclusion is inescapable within his system: their fundamental nature (*ousia*) must be that of a "slave."

In this reading, the "master/slave" dynamic is transposed from the socio-economic realm to the internal structure of the *Psyche*. The "master" is the awakened *Nous*, the Navigator who actively commands their own cognitive faculties. The "slave" is the same individual in a state of epistemological passivity, enslaved to unexamined *Doxa*. While this is almost certainly not what Aristotle intended, it is a coherent extension of his own logic. It stands as the ultimate warning of the synchronic worldview: it mistakes a temporary state of subservience or potential for an immutable essence, with devastating human consequences.[20]

11.5 The NPN Ideal: Society as an Emergent Property of the *Navigator*

When we invert the metaphysical foundation—replacing the static *Demiourgos* (who orders existing matter) with the dynamic *Genetor* (who generates new forms)—the political ideal shifts radically. We move from a vision of the State as a sculptor of citizens to a vision of Society as an emergent ecosystem. In this view, the health of the collective is not measured by its adherence to a rigid blueprint, but by its capacity to foster the *Energeia* of its individual components.

11.5.1 The Navigator Polis: Beyond the Hylomorphic State

In stark contrast, the Neo-Pre-Platonic Naturalist ideal inverts the Aristotelian relationship. Society is not the shaper of the individual, but the

---

20. For the political lineage of the idea, linking it to concepts of arbitrary rule, see Mary Nyquist, *Arbitrary Rule: Slavery, Antislavery, and the Roots of American Freedom* (Chicago: University of Chicago Press, 2013).

emergent, collective expression of the individuals within it.[21] The highest social ideal is not a perfect structure, but a culture composed predominantly of individual Navigators—those who have undergone the awakening to the *Archē* and its *Logos* and who engage with reality *as it is*.

This ideal should not be confused with modern political constructs. Its core value is not liberty or merit, but **navigational capacity** itself. This ideal society, which we must stress is a theoretical horizon and not a practical near-term goal, would be characterized by:

- **The Primacy of the Navigator:** The fundamental unit of society is not the citizen defined by law, but the *Navigator*—an individual whose *Nous* is actively engaged in modeling the *Logos* of the *Archē*.
- **Eudaimonia as the Default State:** *Eudaimonia* is not an achievement orchestrated by the state, but the natural outcome of a life lived in functional alignment with reality.
- **Dynamic Organization through *Philia* and *Neikos*:** Social and economic roles are fluid, self-organizing networks shaped by the gravitational pull of collaboration (*Philia*) and the necessary force of differentiation and creative destruction (*Neikos*).

11.5.2 The Chasm Between Ideal and Practice: The Navigator's Long Game

It is critical to state unequivocally that this ideal is not a political program. It is a **direction**, not a destination. The creation of such a culture is the work of centuries, requiring the sustained, multi-generational leadership of Navigators dedicated not to seizing power, but to the re-education of the human *Psyche*. It necessitates guiding individuals through the stages of *Lysis*, *Aporia*, and *Energeia* until the Navigator becomes the cultural norm. The NPN project, therefore, is inherently a long-term one, focused on shifting the foundational *Navigator Polis* of culture rather than winning the next election. Its political strategy is, first and foremost, pedagogical.

---

21. This view of society as a bottom-up emergent phenomenon aligns with the tradition of spontaneous order. See John Stuart Mill, *On Liberty* (London: John W. Parker and Son, 1859). Mill argues that individual genius and character are the engine of social progress, not products of state design.

11.6 Conclusion: The Harbor and the Ocean

Tracing the arc of Aristotle's thought reveals a consistent and powerful pattern. It is the most comprehensive and coherent system ever devised for understanding a synchronic, intelligible, and teleologically-ordered cosmos. He provided the ultimate philosophy for building and maintaining the **harbor**—a magnificent, static structure where a fixed human nature can find its perfect, pre-ordained rest.

The root of the Aristotelian impasse is not a logical error but a foundational choice: the **flattening** of *Physis* into a synchronic map. This single methodological commitment forced his system to find its ground in a Prime Actualizer, its explanation in a pre-ordained *Telos*, and its politics in the perfection of a fixed essence. He built a closed circle of intelligibility, a magnificent and self-consistent cathedral of thought.

But a harbor, by definition, is a place one leaves to embark on a journey.

Neo-Pre-Platonic Naturalism starts from the opposite shore. Where Aristotle's system finds its ground in a Prime Actualizer, NPN posits the non-negotiable **Primacy of the *Arche***. *Physis* is not a given, ordered whole to be catalogued, but a lawful, dynamic process unfolding within the boundless context of the *Apeiron*. The *Nous* is not the highest activity of a substance, but the emergent Navigator within that process. *Eudaimonia* is not the fulfillment of a pre-written essence, but the successful, ongoing activity of a Navigator whose models are functionally aligned with the *Logos* of the *Archē*.

Aristotle provided the definitive map of a static universe. NPN provides the physics and the principles for navigating a dynamic one. His system is the grand, finished symphony of the harbor. NPN is the theory of music for the open ocean, explaining how such a symphony can be composed, performed, and, when necessary, rewritten by the Navigator who sails into the beautiful ocean of the *Archē*.

Table 11.2 The Great Philosophical Divergence

Tradition	Primary Error	Resulting Dilemma	NPN Resolution
Platonism (Middle Dialogues)	**Transcendent Inversion:** Locates the *Logos* (Forms) outside the *Archē*.	**Mind/Matter Dualism:** Creates an unattainable ideal and pathologizes the physical world.	Grounds the *Nous* as an emergent product of the *Archē* (Mind → Matter).
Aristotelianism	**Synchronic Flattening:** Describes reality as a static hierarchy of fixed essences.	**The *Nous* Aporia:** Cannot account for the dynamic, generative nature of consciousness without a transcendent spark.	Replaces fixed **Being** with dynamic **Becoming** (*Logos* as Becoming).

Chapter 12

The Politics of the Navigator: Philia, Neikos, and the Second-Best State

"The Nocturnal Council... is the anchor of the whole state. Just as the soul and the head are the most important parts of the living creature... so this council must be the intellect of the city, keeping watch over the laws and the safety of the state." – Plato[1]

"He who commands that law should rule may thus be regarded as commanding that God and reason alone should rule; he who commands that a man should rule adds the character of the beast." – Aristotle[2]

12.1 Introduction: The *Polis* as a Natural Phenomenon

Modern political discourse is a cacophony of competing ideologies, each claiming a monopoly on justice. From the NPN vantage, this is a synchronic conflict over static models.[3] The debate between left and right, liberal and conservative, is often a struggle between misplaced *Philia* (an

---

1. Plato, Laws, Book XII, 960d–961d. In the NPN framework, the Nocturnal Council serves as the civilizational analog to the Nous. It is the emergent "Exception Handler" that intervenes when the automated habits of the Logos (tradition and static laws) face the novelty of the Archē. This represents the final evolution of the Navigator: the move from individual Dikaiosynē to the maintenance of order at the scale of the Polis.

2. Aristotle, *Politics*, 1287a28–30. This articulates the NPN principle that the "Second-Best State" must be ruled by impersonal Law (*Logos*), not the arbitrary will of a "beast" (*Thymos*). In *The Complete Works of Aristotle: The Revised Oxford Translation*, ed. Jonathan Barnes (Princeton, NJ: Princeton University Press, 1984).

3. John Stuart Mill, *On Liberty* (London: John W. Parker and Son, 1859), 36. Mill argues that the "collision of opinions" is essential for discovering truth. NPN identifies the

over-emphasis on collective cohesion that stifles the individual) and maladaptive *Neikos* (an over-emphasis on individual strife that shreds the social fabric).

To escape this deadlock, we must execute a Diachronic Turn and ask a more fundamental question. We will not begin by asking, "What is the ideal state?" That is the Platonic error, the *Demiourgos* mode of imposing a blueprint. Instead, we must ask: "What is a society, according to the NPN framework?"

The *polis* is not an artificial construct. It is a natural phenomenon[4], an emergent, large-scale network of *Navigators* and potential *Navigators*, existing within the *Archē* and subject to its impersonal, causal *Logos*.[5]

12.2 The NPN Analysis: The *Physis* of the *Polis*

To navigate the political landscape effectively, we must first strip away the *Doxa* of ideology and view the *Polis* through the lens of naturalism. Politics is not a separate magisterium of moral abstractions; it is the physics of the *Hormē* operating at scale. It is the complex, non-linear result of millions of distinct navigators interacting within the constraints of the *Archē*, driven by the eternal polarity of *Philia* and *Neikos*.

---

modern deadlock as a failure of this collision—a "static *Neikos*" where positions harden rather than evolve.

4. For the Aristotelian view of the *polis* as a natural growth from human sociality, see David Roochnik, "What Is Natural about Aristotle's Polis?," *Polis* 26, no. 2 (2009): 217–40. NPN extends this: the *polis* is the extended phenotype of the Navigator.

5. On Plato's late view of the *polis* and the role of law in shaping character, see Julia Annas, *Virtue and Law in Plato and Beyond* (Oxford: Oxford University Press, 2017), 166–90. F. A. Hayek's concept of "spontaneous order" (*kosmos*) emerging from individual interactions provides a modern correlate. See F. A. Hayek, *Law, Legislation and Liberty*, Vol. 1 (Chicago: University of Chicago Press, 1973), 37–38.

12.2.1 Society as an Emergent Network of Navigators

A society is not a top-down construction but a bottom-up emergence.[6] Its fundamental unit is not the "citizen" as defined by law, but the *navigator*—a conscious agent possessing *Hormē* and the capacity for model-building. The quality of a society is a direct function of the prevalence and efficacy of the Navigators within it. A culture of individuals operating in *Energeia* will produce a radically different *polis* than a culture dominated by unexamined *Doxa*.

12.2.2 The Constitutive Forces: *Philia* and *Neikos* as Social Dynamics

The structure and evolution of this network are governed by the same cosmic forces that shape the rest of the *Archē*:[7]

Philia **(Φιλία) as the Force of Social Cohesion:** This is the impulse behind attraction, union, and integration. In the social sphere, *Philia* manifests as:[8]

- **Trust:** The foundation of all contract and cooperation.
- **Empathy and Affiliation:** The bonds of family, friendship, and tribe.
- **The Social Contract:** The implicit agreement to forego some individual *Neikos* for the greater benefits of collective security and prosperity.
- **Culture and Shared Identity:** The narratives, rituals, and laws that create a unified "we."

---

6. This naturalistic approach to political theory, grounding justice in the basic structure of society rather than transcendent principles, finds a modern analogue in John Rawls, *A Theory of Justice* (Cambridge, MA: Belknap Press, 1971), 3–22. The bottom-up emergence of social order is also central to Hayek, *Law, Legislation and Liberty, Vol. 1*, 35–54.

7. Emile Durkheim, *The Division of Labor in Society* (New York: The Free Press, 1997), 79–80. Durkheim's distinction between "mechanical" and "organic" solidarity parallels the NPN analysis of *Philia* (cohesion) and *Neikos* (differentiation).

8. The NPN analysis of social *Philia* is grounded in evolutionary mechanisms: Kin Selection (Hamilton, 1964) and Reciprocal Altruism (Trivers, 1971). These validate that "Love" is a bottom-up survival strategy, not a top-down command.

***Neikos* (Νεῖκος) as the Force of Social Differentiation:** This is the impulse behind separation, division, and competition. In the social sphere, *Neikos* manifests as:

- **Competition:** For resources, status, and mates.
- **Justice and Accountability:** The necessary force that checks anti-social behavior and holds individuals responsible.
- **Individuation:** The drive that asserts the self against the collective.
- **Creative Destruction:** The friction that breaks down maladaptive or corrupt institutions, making way for new forms.

A healthy society is not one that eliminates *Neikos*, but one that harnesses it. Unchecked *Philia* leads to a stagnant, tyrannical collective where no one is accountable. Unchecked *Neikos* leads to a war of all against all, a Hobbesian nightmare. **Societal *Dikaiosynē* is the functional harmony of these two forces, the social analogue to the individual's psychic *Dikaiosynē*.**

12.2.3 The Telos of the *Polis*: Diachronic Stability

The ultimate purpose, or *Telos*, of a society is not to achieve a static utopia but to ensure its **diachronic stability**—its ability to persist and flourish across time. This is not mere survival, but the capacity to navigate the dynamic, often chaotic processes of *Physis*—economic shifts, environmental challenges, internal conflicts, and external threats—without collapsing. This diachronic stability is the social equivalent of *Eudaimonia*. It is the state of a society in *Energeia*, effectively managing the eternal tension between *Philia* and *Neikos*.

12.3 Two Archetypal Forms: The Necessary Poles of Social Tension

From this analysis, two fundamental, amoral orientations toward political organization emerge. These are not moral categories of "good" and "evil," but functional descriptions of how a society can channel the cosmic forces of *Philia* and *Neikos*. It is vital to understand that the *Demiourgos* and *Genetor polis* are **ideal types.** No real society is purely one or the other;

CHAPTER 12. NPN POLITICS

health is found in the dynamic, context-specific balance of these forces, and the capacity to shift the balance as circumstances demand.[9]

12.3.1 The *Demiourgos* Polis: The Imperative of Order

- **Core Principle:** The state functions as a stabilizing craft. Its drive is to create and maintain order, predictability, and high-fidelity replication of a successful model.

- **Expression of Forces:** It is the primary social instrument of *Philia* as **cohesion and unity**. It builds the walls, writes the laws, and forges the shared identity that protects the group from internal chaos and external threats.

- **Virtues (Its Necessary Function):**

    - **Stability:** Provides the safety and predictability required for long-term projects and complex cooperation.
    - **Efficiency:** Can mobilize resources and coordinate action effectively towards a common goal.
    - **Continuity:** Preserves hard-won knowledge and tradition across generations.

- **Pathology (The Imbalance):** Occurs when the *Demiourgos* function becomes totalizing. It then manifests as:

    - **Stagnation:** The suppression of all *Neikos* (dissent, innovation, competition) leads to a brittle society unable to adapt.
    - **Tyranny:** The blueprint for order becomes an end in itself, crushing the individual *Navigators* it was meant to protect.

A society with no *Demiourgos* principle is pure chaos—a failed state.

---

9. The tension between *Philia* and *Neikos* structures modern political philosophy. Robert Nozick's libertarianism (*Anarchy, State, and Utopia*, 1974) represents a *Neikos*-weighted system (Negative Liberty). Isaiah Berlin's "Positive Liberty" (*Four Essays on Liberty*, 1969) aligns with *Philia*.

12.3.2 The *Genetor* Polis: The Imperative of Adaptation

- **Core Principle:** The state functions as an adaptive network. Its drive is to generate novelty, foster resilience, and cultivate the conditions for new models and *Navigators* to emerge.

- **Expression of Forces:** It is the primary social instrument of *Neikos* as **differentiation and innovation**. It drives the competition of ideas, the breaking of obsolete traditions, and the individuation that allows for creativity.

- **Virtues (Its Necessary Function):**

  - **Resilience:** A society that can innovate and debate is better equipped to handle novel crises.
  - **Flourishing:** Creates the space for individual *Energeia* and the expression of diverse human potentials.
  - **Discovery:** The engine of scientific, artistic, and philosophical progress.

- **Pathology (The Imbalance):** Occurs when the *Genetor* function becomes unmoored. It then manifests as:

  - **Fragmentation:** Unchecked *Neikos* shreds the social fabric of *Philia*, leading to conflict, nihilism, and a loss of shared purpose.
  - **Incoherence:** The constant churn of novelty without a stable core leads to disorientation and an inability to act collectively.

A society with no *Genetor* principle is a sterile, static prison—a museum, not a civilization.

The central political problem is therefore not to choose one, but to structure a society that has a strong *Demiourgos* backbone to provide stability, while protecting *Genetor* spaces to ensure long-term adaptation and flourishing. The tension between them is not a problem to be solved, but the engine of a healthy, diachronically stable society.

CHAPTER 12. NPN POLITICS

Table 12.1 The Political Archetypes: Order vs. Adaptation

Feature	Demiourgos Polis (The State as Craft)	Genetor Polis (The State as Network)
Core Principle	**Stabilization:** Creating order, predictability, and high-fidelity replication.	**Adaptation:** Generating novelty, fostering resilience, and cultivating Navigators.
Dominant Force	**Social *Philia*:** Cohesion, unity, laws, and shared identity.	**Social *Neikos*:** Differentiation, innovation, competition, and debate.
Virtues	Stability, Efficiency, Continuity.	Resilience, Flourishing, Discovery.
Pathology	**Stagnation:** A brittle prison unable to adapt; Tyranny.	**Fragmentation:** Nihilism and incoherence; Chaos.

12.4 The NPN Prescription: The Second-Best State – A Society of Scaffolds

The functional reality of society is that most of the population exists in a state of unexamined *Doxa*, unprepared for the *Lysis* required to become a Navigator. The primary political challenge, therefore, is not to impose an ideal, but to manage this transition—to build a society that is stable enough to function in the present while creating the conditions for more Navigators to emerge in the future. This is the "second-best" state:[10] a **society of scaffolds.**[11]

---

[10]. Plato, *Laws*, 875a–d. This concept of the **"second-best" state ruled by impersonal law** is the practical consequence of Plato's late-period thought, providing a Demiourgos framework for stability when perfectly wise individual Navigators are unavailable (cf. Chapter 9, Section 9.4.2).

[11]. The concept of societal "scaffolds" that enable human flourishing without determining outcomes aligns with Martha Nussbaum's capabilities approach. See Martha C. Nussbaum, *Creating Capabilities: The Human Development Approach* (Cambridge, MA: Belknap Press, 2011), 17–45.

12.4.1 The Necessary *Demiourgos*: Keepers of the Functional Mythos

A population that cannot yet self-govern through the *Somatic logos* of the Navigator must be governed by an external *Logos*. This is the essential, amoral role of the *Demiourgos* in the second-best state. They are not tyrants, but **the keepers of the functional mythos**.[12]

- **The "Mythos" as a Proto-Logos:** These are the shared stories, national identities, ethical codes, and legal principles that provide a simplified, operational model of reality for the populace. They are the "noble lies" or "useful fictions" that foster *Philia* and maintain social cohesion.[13]
- **The *Demiourgos* Function:** The political and cultural institutions—governments, courts, educational systems—must act as *Demiourgoi*: crafting, maintaining, and judiciously updating this mythos. They provide the stable, predictable "blueprint" for society that allows for daily life to function without constant existential crisis.
- **The Goal of the *Demiourgos*:** The ultimate purpose of this top-down structure is not to perpetuate itself, but to create a safe container within which individuals can, over time, be gently guided toward examining their *Doxa*. Its success is measured by how many citizens it helps graduate from needing a mythos to seeking the *Logos*.

12.4.2 Cultivating *Navigators*: The Strategic Goal

While the *Demiourgos* manages the mythos for the many, the state must simultaneously act as a *Genetor* for the few. Its long-term survival depends

---

12. This concept of a hidden, philosophically active core preserving the state's alignment anticipates Plato's **Nocturnal Council** (*Nykterinos Syllogos*) in *Laws* Book XII (961a–968e). Just as the NPN "Keepers" maintain the functional mythos for the "Second-Best State," Plato's Council is charged with knowing the philosophical reality (*Logos*) behind the laws to prevent the state's corruption.

13. For the concept of a "noble lie" used for social cohesion, see Plato, *Republic*, 414b–415d. NPN reinterprets this not as deceit, but as *Heuristic Simplification*—providing a map resolution appropriate to the user.

CHAPTER 12. NPN POLITICS

on strategically increasing the number of *Navigators* within its population.

- **Sanctioned *Aporia* Zones:** The state must deliberately create and protect institutions where the Socratic process is not just allowed, but encouraged. This includes:
    - **Universities and Research Centers:** Protected spaces for radical inquiry and the systematic dismantling of models.
    - **A Free Press and Open Discourse:** A public arena where the *Elenchus* can be applied to the *Doxa* of the day.
    - **Artistic and Cultural Funding:** Supporting the *Genetor* artists who create the simulators for societal *Katharsis*.
- **The *Prohairesis* of the State:** The state itself must make a conscious methodological choice. Its *Demiourgos* institutions (bureaucracy, military, core legal system) should be optimized for stability and predictability. Its *Genetor* institutions (science funding, arts councils, higher education) must be optimized for novelty and discovery. A state that applies a purely *Demiourgos* logic to its *Genetor* sectors will sterilize them.

12.4.3 The Real-World Compass for a Non-Ideal World

The NPN framework provides a diagnostic tool for evaluating the health of any society based on its balance of forces.

A Society in Crisis exhibits:

- ***Demiourgos* Pathology:** Stagnation, censorship, high levels of control, suppression of dissent. (Excessive, maladaptive *Philia*).
- ***Genetor* Pathology:** Radical fragmentation, loss of shared truth, nihilism, inability to coordinate. (Excessive, maladaptive *Neikos*).

A Healthy "Second-Best" State exhibits:

- A strong, impartial *Demiourgos* framework (rule of law) that guarantees basic safety and rights.

- Protected *Genetor* spaces where the *Logos* can be pursued and new models can be generated.
- An educational system that functions as a ladder, offering the stabilizing mythos to those who need it while providing the tools of the Navigator to those who seek them.

The goal of the second-best state is not utopia, but **resilient diachronic stability**. It is a society that is stable enough to not collapse into chaos, but adaptive enough to evolve over time. It understands that its primary resource is not its material wealth, but the cognitive capacity of its people, and it structures itself to cultivate that capacity across generations.

By deducing the necessity of dynamic oscillation from the First Principles of the Archē, T5: The Entropic Mandate provides the ultimate naturalistic refutation of all static utopian ideologies. Any political or social system that attempts to eliminate friction, conflict, and dissent in a pursuit of 'perfect harmony' is unknowingly choosing its own entropic path to failure, because perfect Philia is, in the face of a changing Archē, a guarantee of death. The NPN goal for the Polis is therefore not utopia, but resilient Dikaiosynē (Functional Harmony)—the persistent, energetic balance between the forces of order and change.

12.5 The Ideal: The Navigator Polis: A Society of Harmonious Navigators

The ideal state is not a static blueprint to be achieved, but a dynamic mode of existence to be inhabited.[14] It is the **Navigator Polis**: a society where the *Demiourgos* function has been fully internalized by a citizenry of Navigators, rendering top-down control obsolete. This is not a society without structure, but one where structure emerges organically from the bottom-up, from the aligned *Energeia* of its individuals.[15]

---

14. John Stuart Mill, *On Liberty*, 85. Mill's argument that "the free development of individuality is one of the leading essentials of well-being" aligns with the NPN ideal of a Navigator Polis, where collective flourishing emerges from the symphonic diversity of individuals in *Energeia*.

15. Trivers, "The Evolution of Reciprocal Altruism," 35–57. Reciprocal Altruism is the biological substrate of the "Voluntary Association" that defines the Navigator Polis.

12.5.1 The Constitutive Principle: Internalized *Logos*

In the Navigator Polis, the external scaffolding of the "second-best" state has fallen away because it is no longer needed. The citizen does not follow the law out of fear of punishment (*Neikos*) or blind allegiance to a mythos (*Philia*), but because their own Somatic *logos*—their hard-won, rational understanding of *Physis*—is in perfect harmony with the society's operational principles.

- **Law as Discovered, Not Decreed:** The "laws" of such a society are not commands but the codified discoveries of its Navigators. They are the shared, tested models that have proven most effective for sustaining collective flourishing. Legislation becomes a continuous process of collective inquiry and model-updating, a societal *Elenchus*.
- **The Dissolution of the Ruler/Ruled Dichotomy:** When every individual is a Navigator, governance ceases to be a specialized profession of "keepers." It becomes a distributed function. Leadership rotates based on competence and context, not permanent authority. The Philosopher-King is not a person, but a mode of engagement available to all.

12.5.2 The Organic Division of Labor: Voluntary Specialization in a Navigator Network

A common fear of any "ideal society" is that it demands uniform excellence, creating a homogenous population of philosopher-kings. The NPN ideal presents the opposite: it is the fullest flowering of natural human diversity, because it allows for voluntary specialization based on innate aptitude and discovered passion, without the distortions of coercion or unexamined *Doxa*.

In the Navigator Polis, the varied expressions of the *Hormē* are not suppressed but celebrated and channeled. The societal roles are not assigned but emerge organically from the interplay of individual natures:

- The individual with a potent *Thymos* and a talent for strategic *Logos* naturally gravitates toward leadership and coordination. They govern not as a ruler, but as a facilitator whose authority is voluntarily granted by the group because they demonstrably enhance collective efficacy.
- The individual with a deep *Orexis* for physical *Hylē* and a *Demiourgos* mindset finds fulfillment in the masterful craft of building, engineering, and maintaining the physical infrastructure—the literal scaffolds of society.
- The pure *Genetor*, driven by a need to synthesize new models, flourishes as the artist, theoretical scientist, or philosopher, pushing the boundaries of the collective *Nous*.
- Countless other niches—the caregiver, the teacher, the organizer—are filled by those whose Somatic *logos* finds its highest expression in those roles.

This is not a threat to power, but its transformation. "Power" ceases to be a scarce resource to be hoarded and protected. It becomes a functional capacity that is voluntarily delegated.

- **The Highest Form of Control is Voluntary Consent:** In this system, the "leader class" is not a protected elite. It is a fluid group of individuals empowered by the conscious, rational consent of their peers. Their authority is derived from their demonstrated competence and their alignment with the shared *Logos* of the *polis*.
- **The Dissolution of Coercive Overhead:** This voluntary delegation is the most efficient control system imaginable. The immense, costly apparatus required to maintain power over people—propaganda, surveillance, secret police, ideological enforcement—would be radically minimized, re-purposed, or would wither away, its energy redirected toward empowering the collective project. The society sheds the metabolic burden of internal coercion.

In the Navigator Polis, the harmony is not one of sameness, but of "**symphonic diversity**". Each individual, by following their own unique navigational compass toward *Energeia*, automatically fills the role for which they are best suited, contributing to a whole that is far greater, more resilient, and more adaptive than the sum of its parts. The *polis* understands that the builder, the leader, and the artist are all equally essential Navigators, each mastering a different dimension of *Physis*.

12.5.3 The Dynamics of a Navigator Network: *Philia* and *Neikos* in Harmony

The cosmic forces are not eliminated; they are elevated to their highest, most productive expression.

- *Philia* **as Conscious Symbiosis:** Attraction is no longer based on tribal mythos but on the recognition of shared participation in the *Logos*. Cooperation is deep, voluntary, and emerges from the understanding that collective *Eudaimonia* amplifies individual *Eudaimonia*. This is *Philia* as the gravitational force binding a constellation of self-aware stars.
- *Neikos* **as Creative Friction:** Strife is transformed from destructive conflict into the engine of refinement. It is the necessary and welcomed friction of debate, the competition of ideas, and the process of testing models against reality. In a society of Navigators, *Neikos* is not a threat to personal identity but the very tool for achieving greater alignment with the *Archē*. It is the force that ensures the society remains a learning, adapting system, preventing the stagnation that doomed the rigid *Demiourgos polis*.

12.5.4 The Individual in the Navigator Polis: The Navigator in *Energeia*

The ultimate output of this society is the individual living in a state of sustained *Energeia*.

- **Minimized Internal Conflict:** The cognitive dissonance and psychic tension that plague the unexamined life are largely absent. The Navigator's *Logistikon* is in harmonious command of *Orexis* and *Thymos*; their actions are aligned with their understanding, and their understanding is constantly refined through interaction with a reality they are equipped to handle.
- **The Generative State:** This internal harmony unleashes immense creative potential. Freed from the energy drain of internal conflict and the struggle against oppressive external structures, the individual's *Hormē* finds its fullest expression in generative acts—in art, science, philosophy, and the daily practice of building a better world. A society of such individuals is, by definition, a "happier, more productive" one, because happiness is the subjective experience of functional alignment, and productivity is its natural output.

12.6 Conclusion: The North Star

The Navigator Polis is the **North Star** for the NPN political project—a direction, not a destination.[16] It is the logical culmination of the system's First Principles: if the *Nous* is an emergent navigator, then the best possible society is one composed of such Navigators, consciously cooperating.

This ideal reveals the ultimate purpose of the "second-best" state: to make itself obsolete. The goal of the keepers of the mythos, the builders of scaffolds, and the cultivators of *Navigators* is to work towards a future where their function is no longer needed because every citizen has learned to

---

16. On the ethical implications of Plato's mature political vision in the *Laws*, which moves from ideal theory to a practical framework of law, see Annas, *Virtue and Law in Plato and Beyond*, 136–65. NPN validates Annas's reading: the *Laws* is not a retreat, but the necessary blueprint for navigating a non-ideal world.

keep, build, and navigate for themselves. The journey from the second-best state to the ideal is not a political revolution, but a quiet, relentless, multi-generational awakening—the collective *Lysis* of societal *Doxa* and the emergence of a people who have learned to think for themselves, together.[17]

---

17. The political consequences of epistemological frameworks—how our models of knowledge shape our political arrangements—is a central theme in Hilary Putnam's work. See Hilary Putnam, *Reason, Truth and History* (Cambridge: Cambridge University Press, 1981), 49–74, 103–26.

Chapter 13

The Navigator's Art: An NPN Aesthetics

"Art is the lie that enables us to realize the truth."
– Pablo Picasso[1]

"Poetry is something more philosophic and of graver import than history, since its statements are of the nature rather of universals, whereas those of history are singulars."
– Aristotle[2]

13.1 Introduction: Art as a Natural Human Behavior

Aesthetics has long been the most elusive branch of philosophy, often relegated to the realm of subjective taste, emotional expression, and ineffable emotion. From the NPN vantage, this is a profound misunderstanding of a fundamental human technology.[3] We begin not by asking "What is beautiful?" or "What is good art?" but with a more foundational, naturalistic question: **What *is* art, and why does this seemingly universal behavior exist within the *Archē*?**

---

1. Attributed to Pablo Picasso. NPN interprets this "lie" as a **High-Fidelity Simulation**: a construct that is factually false (the lie) but structurally accurate (the truth) regarding the causal *Logos*.

2. Aristotle, *Poetics*, 1451b5–7. NPN adopts this distinction: History records the entropic flux of the *Archē* (singulars), while Art models the invariant *Logos* (universals). In *The Complete Works of Aristotle: The Revised Oxford Translation*, ed. Jonathan Barnes (Princeton, NJ: Princeton University Press, 1984).

3. This view of art as a fundamental, evolved behavior aligns with a growing body of anthropological thought. See Ellen Dissanayake, *Homo Aestheticus: Where Art Comes From and Why* (New York: The Free Press, 1992). Dissanayake argues art is "making special"—a survival behavior, not a luxury.

The answer, revealed through the NPN lens, is that art is not a luxury or a mere reflection of culture. It is a core epistemic and psychological technology for the Navigator. It is a high-fidelity simulator for the *Psyche*, a mechanism for controlled cognitive crisis, and a society's primary tool for cultural navigation and renewal. This chapter will analyze the *Physis* of art through the NPN framework and derive its indispensable role in the project of human flourishing.

13.2 The Functional Purpose of Art: *Mimesis* as High-Fidelity Simulation

Aristotle identified *mimesis*—imitation—as the defining characteristic of poetry.[4] NPN refines this into a precise technical function: art is the construction of a **high-fidelity simulation**.[5]

A tragic drama, a novel, or a symphony is not a simple "copy" of reality. It is a compressed, controlled model that runs a specific scenario through the lawful *Logos* of the *Archē* as it applies to human experience. It abstracts away the noise of daily life to isolate and operate the core causal chains of character, choice, and consequence.

The Tragic Plot as a Causal Simulator: A well-constructed narrative like *Oedipus Rex* or *Macbeth* models the inescapable workings of *Dikē*. It allows the *Nous* to observe the outcomes of *hamartia* (critical error), the complex interplay of fate and agency, and the rebalancing justice of the cosmic framework, all within a consequence-free environment. It is a flight simulator for the *Psyche*, training the *Logistikon* in pattern recognition and long-term forecasting within the human domain.

---

4. Aristotle, *Poetics*, 1447a13–16.

5. The NPN concept of art as a simulator finds a robust parallel in modern cognitive science, which understands the brain as a prediction engine. See Lawrence W. Barsalou, "Perceptual Symbol Systems," *Behavioral and Brain Sciences* 22, no. 4 (1999): 577–660. Art externalizes the *Logistikon's* modeling process.

13.2.1 The Cognitive Process of Art: *Katharsis* as a Controlled *Lysis*

If art is a simulator, then its emotional power—what Aristotle termed *katharsis* (purgation)—is the psychological result of a controlled, psycho-epistemological crisis. Art is a machine for inducing a safe *Lysis*.[6]

The process mirrors the Socratic method, but operates on the integrated *Psyche* (*Orexis*, *Thymos*, *Logistikon*, and *Nous*) through emotion and narrative, not just logic:

1. **Inducing Productive *Aporia*:** A powerful artwork systematically dismantles the audience's naive or rigid models—*Doxa*—of justice, love, society, or the self. It guides them to a point where their existing cognitive-emotional structures fail, inducing a temporary, safe state of *Aporia*. The world no longer makes sense as they thought it did.
2. ***Katharsis* as Cognitive *Lysis* (Release):** The emotional purgation is the *Lysis*—the release and recalibration that follows this model failure. The intense pity and fear are the somatic signals of the old model breaking down. This purges the erroneous emotional charges and flawed cognitive maps, leaving the *Psyche* cleansed and restructured.
3. **The Outcome:** The audience departs not just "moved," but with a more nuanced, resilient, and accurate internal model of reality. They have, in a profound sense, updated their map. They have survived a simulated catastrophe and are better prepared for the real ones. This is the Navigator honing its skills.

---

6. Aristotle, *Poetics*, 1449b24–28. NPN supports the cognitive interpretation of *katharsis*—it is not merely medical purgation, but ethical clarification. See Jonathan Lear, "Katharsis," *Phronesis* 33, no. 3 (1988): 297–326.

13.2.2 The Artist's *Prohairesis*: *Demiourgos* and *Genetor* modes in Creation

The act of creation reflects the fundamental methodological choice of the *Nous*, a *Prohairesis* between two archetypal orientations. The greatest artists are not permanent residents of one mode but are masters of transition, knowing when to perfect a form (*Demiourgos*) and when to shatter it to create a new one (*Genetor*). The modes are a description of creative actions, not a taxonomy of creative souls:

- ***Demiourgos* Artist (The Master Craftsman):** This artist's *Prohairesis* is one of Structural Analysis and perfect execution within an established tradition. They are master navigators using well-known maps, aiming to achieve a powerful, predictable effect by perfectly applying the *Logos* of a given form. A composer writing a perfect fugue, a poet mastering the sonnet, or a painter working within a strict canon exemplifies this mode. Their goal is **mastery and perfection of the known.**

- ***Genetor* Artist (The Pioneer):** This artist's *Prohairesis* is one of Synthesis. They do not merely follow forms but generate new models of reality. They are cartographers of uncharted psychic and perceptual territory. The inventor of the novel, the pioneer of abstract expressionism, or the composer who forges a new harmonic language is a *Genetor*. They expand the very capacity of the collective *Nous* to apprehend the *Archē*. Their goal is the **discovery of the new.**[7]

Both are essential. The *Demiourgos* provides the stable forms and technical mastery that become the foundation for the *Genetor*'s next leap. A vibrant culture requires both.

---

7. The function of the *Genetor* artist operates analogously to the revolutionary scientist who shifts a prevailing paradigm. See Thomas S. Kuhn, *The Structure of Scientific Revolutions*, 2nd ed. (Chicago: University of Chicago Press, 1970).

Table 13.1 The Artist's Prohairesis: Two Modes of Creation

Feature	Demiourgos Artist (Master Craftsman)	Genetor Artist (The Pioneer)
Goal	Mastery: Perfection of known forms and execution.	Discovery: Synthesis of new paradigms and models.
Method	Structural Analysis within an established tradition.	Mapping uncharted psychic or perceptual territory.
Examples	The perfect Fugue; The Sonnet; Canon painting.	The Novel; Abstract Expressionism; New harmonics.
Social Function	Stabilizes culture (Social *Philia*); Reinforces identity.	Challenges *Doxa* (Cognitive *Neikos*); Triggers adaptation.

13.2.3 The Social Function: Art as a Cultural Feedback Mechanism

Beyond individual training, art functions as a society's self-correcting mechanism and cognitive immune system. It is a culturally sanctioned Navigator that performs an *Elenchus* on the collective *Doxa*.

- **Challenging the *Demiourgos*:** Every society develops a *Demiourgos* tendency—a set of dominant myths, narratives, and power structures that seek to preserve themselves. Art, particularly from the *Genetor* mode, constantly tests and questions these prevailing stories. It fights the pathological tendency to fossilize a single narrative, thereby preventing the societal decay that comes from misalignment with a changing reality.
- **Reinforcing *Philia*:** Shared artistic experiences—from ancient Greek tragedies to modern film—create common reference points, build empathy, and forge the bonds of collective identity. They

are a primary technology for building the social *Philia* that is the bedrock of a cohesive *Polis*.

- **Performing the Societal *Elenchus*:** A powerful modern example is Harriet Beecher Stowe's *Uncle Tom's Cabin*. The novel functioned as a nationwide *Elenchus* for the American *Doxa* of slavery. It did not merely present an argument; it constructed a devastatingly high-fidelity simulation that forced readers—particularly in the North—to confront the internal contradictions of a nation that proclaimed liberty while upholding a system of brutal, familial destruction. By making the abstract horror of the slave system visceral and personal, the novel exposed the fatal flaw in the prevailing societal model, inducing a moral and cognitive *Aporia* on a massive scale. This artistic *Lysis* of a foundational national *Doxa* was a catalytic event in the lead-up to the Civil War, demonstrating art's power to act as a society's conscience and a trigger for structural *Neikos*.[8]

13.2.4 The NPN Prescription: The Pragmatic Imperative of Art

From this analysis of what art *is*, a powerful prescription for its role naturally follows. In a non-ideal world, art is not a luxury. It is a critical piece of societal infrastructure.

- **Art as a Navigational Instrument:** A society that understands NPN would recognize art as it recognizes its scientific laboratories and navigation schools—as an essential institution for its long-term adaptation and flourishing. It is a public good.
- **A Moral and Strategic Simulator:** For the leaders, art models the societal collapse that follows from unchecked *Neikos* and the functional strength derived from cooperative *Philia*. For the populace, it is the primary technology for processing collective trauma and building shared meaning.

---

8. For a historical analysis of the novel's immense cultural impact as a "Social Elenchus," see David S. Reynolds, *Mightier Than the Sword: Uncle Tom's Cabin and the Battle for America* (New York: W. W. Norton & Company, 2011).

- **Cultivating the *Genetor*:** A wise cultural policy would consciously protect and fund its *Genetor* artists, not just its *Demiourgos* artisans. It needs the pioneers who can map the unknown territories of the future human condition.

13.3 The NPN Theory of Beauty: Aesthetic Judgment as a Functional Heuristic

The experience of beauty, often considered the most subjective and ineffable of human feelings, is from the NPN vantage a **deeply practical and evolved technology of the *Psyche*.**[9] It is a rapid, intuitive heuristic of the *Somatic logos* for assessing functional excellence and high potential for flourishing. This assessment operates across two primary domains: the biological and the artistic.

13.3.1 Biological Beauty: The Assessment of *Hormē* and *Philia*

In a potential mate, "beauty" is the perceptual shorthand for a bundle of traits that signal a high probability of successful collaboration (*Philia*) and the effective fulfillment of the constitutive *Hormē*. This is not a conscious calculation but a lightning-fast, valenced assessment performed by the deep strata of *Orexis* and *Thymos*.

- **Symmetry:** A heuristic for developmental stability and low pathogenic load (minimized internal *Neikos*).[10]
- **Vital Signs (clear skin, bright eyes):** A heuristic for current health and robust metabolic function (a system in *Energeia*).[11]
- **Specific Proportions:** Heuristics for fertility, health, and the physical capacity to successfully rear/protect offspring.[12]

---

9. This functionalist account resonates with evolutionary aesthetics, which posits that aesthetic instincts are biological adaptations. See Denis Dutton, *The Art Instinct: Beauty, Pleasure, and Human Evolution* (New York: Bloomsbury Press, 2009).

10. See Karl Grammer and Randy Thornhill, "Human facial attractiveness and sexual selection," *Journal of Comparative Psychology* 108, no. 3 (1994): 233–42.

11. See Bernhard Fink and Nick Neave, "The biology of facial beauty," *International Journal of Cosmetic Science* 27, no. 6 (2005): 317–25.

12. For female WHR: Devendra Singh, "Adaptive significance of female physical attractiveness," *JPSP* 65, no. 2 (1993). For male SHR: Susan M. Hughes and Gordon G. Gallup,

The positive feeling of "beauty" is the reward signal for identifying a candidate who represents a high likelihood of fulfilling the *Hormē* through reproductive *Philia*.

13.3.2 Artistic & Aesthetic Beauty: The Simulated Assessment of *Dikaiosynē*

This biological heuristic is co-opted and sublimated by the *Nous*, applying the same functional assessment to artifacts, landscapes, and abstract forms. We find things beautiful that model or promise states of functional excellence and alignment.

- **A Beautiful Landscape (a fertile valley, a clean water source):** Heuristic for an environment conducive to survival and flourishing.
- **A Beautiful Tool or Machine:** Heuristic for perfect *Demiourgos* functionality, a thing perfectly fulfilling its purpose (*Aretē*).
- **A Beautiful Narrative or Artwork:** A simulator for *Dikaiosynē* (the harmonious *Psyche*) or the successful navigation of life's fundamental tensions.

13.3.3 The Compass of the *Hormē*: Beauty and Ugliness as Navigational Signals

The perception of beauty and its opposite, ugliness, is not arbitrary cultural inventions but the emotional correlates of the *Psyche*'s deepest navigational instincts. **Ugliness** is the negative valance assigned by the *Somatic logos* to signifiers of dysfunction, disease, decay, and maladaptation—the failure of the *Hormē*. **Beauty** is the positive valance assigned to signifiers of its fulfillment.[13]

While cultural *Doxa* can certainly layer secondary associations and symbolic meanings onto these primary somatic heuristics—teaching us to find

---

"Sex differences in morphological predictors," *Evolution and Human Behavior* 24, no. 3 (2003).

13. See Dutton, *The Art Instinct*, chaps. 2, 5. Beauty acts as a positive reinforcement signal for behaviors and environments that serve the organism's fitness.

CHAPTER 13. NPN AESTHETICS

specific styles or artifacts beautiful—the core engine of the aesthetic response is this non-negotiable, biological assessment of functional potential. A culture cannot, in the long run, successfully condition its members to find the smell of rot fundamentally beautiful or the sight of vibrant health inherently repulsive, because such a reversal would directly sabotage the *Hormē*'s core imperative.

The NPN Definition of Beauty: Beauty is the positive valance (the "good" feeling) assigned by the *Psyche*'s evolved heuristics upon perceiving a signifier of high functionality, health, or potential for flourishing—whether in a mate, an object, a system, or a simulated model of reality.

This grounds aesthetics in the non-negotiable logic of the *Archē*. Beauty is the *Hormē*'s taste for the conditions of its own success. It is the ancient, somatic compass that points the Navigator, often without its conscious knowledge, toward what is likely to be *good for it*, guiding its journey toward *Eudaimonia* through an intuitive grasp of the *Logos* of life.

13.4 Conclusion: The Final Affirmation

We began this inquiry in the silence of the *Apeiron*, confronting a universe that offers no pre-written scripts and no guaranteed safe harbor. We have traced the emergence of the *Nous*, the struggle for *Epistēmē*, and the construction of the *Navigator Polis*. Yet, logic alone is not enough to sustain the human spirit against the vast indifference of the cosmos.

Aristotle gave us the mechanics of Art, but Neo-Pre-Platonic Naturalism reveals its soul. Art is not merely a simulation or a safe space for failure; it is the Navigator's supreme act of defiance against Entropy. It is the audacity to take the raw, chaotic *Hylē* of existence and stamp it with the seal of the *Logos*, creating form, beauty, and meaning where none existed before.

If Philosophy provides the map, Art provides the courage to walk the territory. It allows us to look into the abyss of the *Neikos*—the tragedy, the suffering, the loss—and not turn away in despair, but transmute it into the gold of *Katharsis*. To live the unexamined life is to sleepwalk; to live the examined life is to wake up; but to live the **Artistic** life is to dance with the *Physis* that sustains us.

The Navigator, in the end, is an artist of the real. We do not escape to a world of Forms; we stand firmly in this one, with all its danger and beauty, and we build. This is the only immortality available to us, and the only *Eudaimonia* worth seeking: to be, for a brief and shining moment, a conscious, generative focal point of the universe understanding itself.

Appendix A

The General Zero Principle: Avoiding the Infinite Regress of Determination

"If we are not to embark on an infinite regress, we must arrive at some first principle which is not derived from anything else."
— Aristotle[1]

A.1 The Problem of Infinite Regress

Every philosophical system faces a fundamental challenge: how to ground determination without falling into either infinite regress or circular reasoning. Consider the question of identity:

1. If we say "A is defined by not being B," we must then ask: "What defines B?"
2. If B is defined by not being C, and C by not being D, we enter **infinite regress**.
3. If we eventually say "Z is defined by not being A," we have **circular reasoning** — a closed loop that provides no ultimate foundation.

This problem manifests at every level:

- **Metaphysical:** What grounds the existence of particular things?
- **Epistemological:** What justifies knowledge claims?
- **Semantic:** What gives words meaning?

---

1. Aristotle, *Posterior Analytics* I.3, 72b5–10. NPN adopts this axiom but rejects Aristotle's solution (the Unmoved Mover) in favor of the *Apeironic Context*. In *The Complete Works of Aristotle: The Revised Oxford Translation*, ed. Jonathan Barnes (Princeton, NJ: Princeton University Press, 1984), 1:118.

A.2 Circular Reasoning as Infinite Regress

Circular reasoning is not fundamentally different from infinite regress — it is **infinite regress compressed into a loop**. When A depends on B and B depends on A, we have:

$$A \to B \to A \to B \to A \to \cdots$$

This is equivalent to the infinite sequence:

$$A \to B \to A' \to B' \to A'' \to B'' \to \cdots$$

Where each iteration provides no more foundation than the last. The loop offers only the *illusion* of closure.

A.3 The GZP Solution: The Indeterminate Foundation

The General Zero Principle provides the necessary escape from this dilemma:

General Zero Principle

GZP: GENERAL ZERO PRINCIPLE

Statement: For anything to possess determinate identity, meaning, or existence, it must exist within a delimited context set against an **indeterminate background**.

Justification: Infinite regress is avoided only by positing a ground that is not a "thing" among things. To define "A" requires "not-A". If "not-A" is also determinate ("B"), the regress continues. It must stop at the indeterminate.

Corollary: Zero Principle: The ultimate foundation cannot itself be determinate, for then it would require further foundation. It must be **indeterminate**.

A.3.1 The Two Levels of GZP

The principle operates at two distinct but related levels:

- **Transcendental GZP (General):** All determination requires an indeterminate ground. This is the **condition for any determination whatsoever**—whether a physical object, a concept, or a state of being.

- **Ontological GZP (Zero Principle, ZP):** System identity requires an indeterminate complement. This is the **specific application to bounded systems**—the *Archē* requires the *Apeiron*, a particular requires its Form (in Plato's system), a thing requires its context.

ZP is GZP applied to the specific case of *system identity*. The first is transcendental; the second is ontological.

A.3.2 How GZP Resolves the Regress

For Physical Objects: A tree is bounded space against **unbounded space** (the void).

- Not: "Tree is not-rock, rock is not-water, water is not-air..." (regress)
- But: "Tree is bounded form against formlessness" (foundation in the indeterminate)

For Concepts: "Justice" is bounded meaning against **unbounded semantic field**.

- Not: "Justice is not-injustice, injustice is not-tyranny, tyranny is not-anarchy..." (regress)
- But: "Justice is delimited concept against conceptual indeterminacy"

For the Cosmos: The *Archē* is bounded reality against **unbounded potential** (*Apeiron*).

- Not: "Our universe exists in a multiverse, which exists in a megaverse..." (infinite regress)
- But: "Determinate reality exists against indeterminate ground"

A.4 Historical Recognition of the Problem

The General Zero Principle did not emerge in a philosophical vacuum. Its logical necessity has been recognized, in various forms and with varying degrees of clarity, throughout the history of Western thought. The following sections trace this recognition from its first intuitive glimpse in Anaximander to its modern formalizations.

A.4.1 Anaximander's Insight

Anaximander recognized that no determinate element (water, air, fire) could serve as ultimate principle, for each would require something beyond itself to define it. His *Apeiron* was the first explicit acknowledgment that **the foundation must be indeterminate**.

A.4.2 Plato's Struggle and the Third Man Regress

Plato's Theory of Forms attempts to escape regress by positing perfect, self-defining Forms. But this merely shifts the problem: if Forms are determinate, they too require grounding. The "Third Man Argument" in the *Parmenides* demonstrates that this leads to infinite regress.[2] Plato's late dialogues represent his recognition that the ground of determination cannot itself be fully determinate.

A.4.3 Aristotle's Formal Solution and Its Limit

Aristotle explicitly states that to avoid infinite regress, we must posit an **unmoved mover**—a first principle that is not itself moved or defined by

---

2. Plato, *Parmenides*, 132a–133a. NPN interprets this as Plato's own admission that a determinate ground (Forms) cannot solve the regress problem. In *Plato: Complete Works*, ed. John M. Cooper (Indianapolis, IN: Hackett Publishing Company, 1997).

APPENDIX A. THE GENERAL ZERO PRINCIPLE

anything else.[3] His potentiality/actuality distinction provides a framework for understanding determination within the *Archē*, but when applied to the ultimate ground itself, it risks committing the very error it seeks to avoid: characterizing the indeterminate in determinate terms.

Aristotle's unmoved mover as *pure actuality* (with no potential) becomes itself a metaphysical posit that resists its own logic—a "something" that paradoxically has no "outside." GZP, in contrast, does not attempt to characterize the ground positively. It only stipulates the **relational necessity**: that for any determinate system to exist, there must be an indeterminate complement. The *Apeiron* is not "pure potentiality" or "pure actuality"—both are determinate concepts. It is simply **that which lies beyond the boundary of the *Archē***, the necessary "outside" whose nature we cannot model without violating FP5 (Impotence Before the *Apeiron*).

A.4.4 Cross-Cultural Validation: The Universal Logic of Wuji

The necessity of an indeterminate ground is not a uniquely Western discovery; it is a structural feature of human thought grappling with reality. Ancient Chinese philosophy independently identified the "Nameless" (*Wuji* or "Limitless") as the necessary precursor to the "Named" (*Taiji* or "Supreme Polarity").

> "*The Nameless is the beginning of Heaven and Earth.*" — Laozi[4]

Just as the *Archē* requires the *Apeiron*, the *Taiji* (determinate reality) requires *Wuji* (indeterminate void). This convergence confirms that GZP is not a cultural artifact, but a logical necessity for any system of identity. (For a complete analysis of the structural isomorphism between NPN and Daoism, see **Appendix H**).

---

3. Aristotle, *Metaphysics* XII.7, 1072a20–25. Aristotle attempts to solve the regress with "Pure Actuality," but NPN argues this violates the GZP by positing a determination without an indeterminacy.

4. Laozi, *Dao De Jing*, trans. Wing-Tsit Chan, in *A Source Book in Chinese Philosophy* (Princeton: Princeton University Press, 1963), 139 (Chapter 1). This confirms GZP as a trans-cultural logical necessity, not a parochial Western invention.

A.5 Modern Formalization: From Spinoza to Spencer-Brown

While the ancients intuited the necessity of the indeterminate, modern philosophy and mathematics undertook the task of defining its logical boundaries. The progression from Spinoza to Gödel represents the rigorous formalization of the Zero Principle, moving it from metaphysical assertion to mathematical proof.

A.5.1 Spinoza: Determination as Negation

The logic of the GZP finds its philosophical crystallization in Spinoza's *"Determinatio est negatio"*.

Spinoza's Insight: Two centuries before Spencer-Brown, Baruch Spinoza articulated the core logical insight in his correspondence: *"Determinatio est negatio"* (Determination is negation).[5] For Spinoza, to determine or define a thing is to limit it, which necessarily implies a negation of the infinite or unbounded background. In NPN terms, Spinoza recognized that the *Archē* (the world of determinate modes) is only possible through the stipulative negation of the *Apeiron* (the infinite substance, or *Deus sive Natura*).

A.5.2 Gödel's Incompleteness: The Formal Proof of GZP

Kurt Gödel's Incompleteness Theorems (1931) provide the rigorous mathematical demonstration of what GZP states philosophically: **no sufficiently complex determinate system can ground itself.**[6]

First Incompleteness Theorem: Any consistent formal system capable of expressing basic arithmetic contains true statements that cannot be proven within the system.

---

5. Benedict de Spinoza, "Letter 50," in *Complete Works*, ed. Michael L. Morgan, trans. Samuel Shirley (Indianapolis: Hackett Publishing, 2002), 892. Spinoza correctly identifies that to be *this* is to not be *that*, requiring an infinite field of "not-that" (Substance) to carve from.

6. Kurt Gödel, "On Formally Undecidable Propositions of *Principia Mathematica* and Related Systems I" (1931). Gödel proved that any formal system powerful enough for arithmetic is necessarily incomplete—it contains truths it cannot prove.

APPENDIX A. THE GENERAL ZERO PRINCIPLE

Second Incompleteness Theorem: No consistent system can prove its own consistency from within its own axioms.

The GZP Reading: Gödel demonstrated that the *ground of determination* (the axioms and rules of inference) cannot themselves be *determined* (proven consistent) from within. The system requires an indeterminate complement—either: - An external meta-system (which faces the same problem—infinite regress) - Acceptance of an unprovable ground (the indeterminate foundation)

Hilbert's Program sought to eliminate this vulnerability by grounding all mathematics in complete, consistent formal axioms—a quest for absolute, self-contained determination.[7] Gödel showed this is **logically impossible**—not because mathematics is flawed, but because GZP is a structural necessity of any sufficiently expressive formal system.[8]

This isn't mathematics "failing." It's mathematics **discovering the same boundary** Anaximander intuited 2,500 years earlier: determination requires the indeterminate. Where philosophy identified this through metaphysical reasoning, Gödel proved it through formal logic. The Incompleteness Theorems are not a limitation of mathematics but a confirmation of the deep structure of reality itself.[9]

A.5.3 Spencer-Brown: The Calculus of Distinction

Spencer-Brown's Calculus: George Spencer-Brown's *Laws of Form* provides the formal mathematical system that corresponds to this insight.

---

[7]. David Hilbert, *Grundlagen der Geometrie* (Leipzig: B. G. Teubner, 1899). Hilbert represents the last great attempt at the *Demiourgos* dream of total, self-contained order.

[8]. For an accessible exposition, see Ernest Nagel and James R. Newman, *Gödel's Proof*, revised ed. (New York: New York University Press, 2001).

[9]. For analysis of misapplications vs. proper scope, see Torkel Franzén, *Gödel's Theorem: An Incomplete Guide to Its Use and Abuse* (Wellesley, MA: A K Peters, 2005).

His calculus begins with a single, primordial injunction: "Draw a distinction."[10] He demonstrates that the "unmarked state" is the necessary background for any "mark" (value, existence, distinction) to appear. The first distinction creates a universe by separating an "inside" from an "outside."

In NPN terms, Spencer-Brown provides the logical proof that the *Archē* (the Marked State, the world of forms) cannot exist without the *Apeiron* (the Unmarked State). The act of distinction is the formal equivalent of Anaximander's *Adikia*—the primordial "cut" that generates bounded existence from the boundless. Spinoza gave the principle its philosophical form; Spencer-Brown gave it its mathematical form.

A.6 The *Apeiron* and the Meta-Regress

A final objection arises: if *everything* requires an indeterminate background, does the *Apeiron* itself require one? The answer is that the *Apeiron* is **not a "thing"** in the determinate sense. It is the **condition for things**, not another thing among things. To ask "what grounds the *Apeiron*?" is to commit a category error—to apply the logic of determination to the very ground that makes determination possible. Within our *Archē*, we cannot know whether there is another *Archē* outside it, bounded by its own *Apeiron*. That question points squarely to the *Apeiron* itself—the domain where empirical observation is impossible and no logical operation can be grounded. This is why the NPN system is honest: everything, including this foundational claim, resides on the **Confidence Gradient**. We adopt GZP not as dogmatic certainty, but as the highest-confidence model that stops the regress without contradiction.

A.7 Application to NPN First Principles

The power of the General Zero Principle is demonstrated by how it grounds and clarifies the entire NPN system. Where other metaphysical

---

10. George Spencer-Brown, *Laws of Form* (London: George Allen & Unwin, 1969), 3. Spencer-Brown mathematically formalizes the *Zero Principle*: the "Mark" (Identity) is impossible without the "Unmarked State" (Void).

frameworks risk circularity or infinite regress, GZP provides the clean, non-circular foundation that makes systematic coherence possible.

A.7.1 FP3: The *Logos* and Asymmetrical Polarity

The original formulation of FP3 risked circularity: *Philia* defines *Neikos* and *Neikos* defines *Philia*. Without an external logical anchor, this mutual definition could appear as a closed loop. The **General Zero Principle (GZP)** provides the universal logical warrant, and the **Zero Principle (ZP)** applies it to system identity.

1. **Universal Law (GZP):** Any determinate concept or entity requires an indeterminate background to possess meaning or existence.
2. **System Application (ZP):** For a specific system (Figure) to possess identity, it must emerge from a contrasting field (Ground).
3. **The Specific Cases:**

- **Macro-Cosmic:** The *Archē* (The Totality of Things) is determinate; therefore, by ZP, it requires the *Apeiron* (The Infinite/Indeterminate) as its background.
- **Micro-Dynamic (FP3):** Within the *Archē*, any formed entity (*Philia*) is determinate; therefore, by ZP, it requires a background of separation and dissolution (*Neikos*) to be distinct.

1. GZP (Universal Logic) : Determinacy $\Rightarrow \exists$ Indeterminate Background
2. ZP (Object Necessity) : Figure $\Rightarrow \exists$ Ground
3. FP3 (Dynamic Realization) : $\underbrace{Philia \text{ (Integration)}}_{\text{Figure}} \Rightarrow \underbrace{\exists Neikos \text{ (Entropy)}}_{\text{Ground}}$

Thus, the polarity is not a circular "war" between equals, but a hierarchical derivation of identity:

No regress. No circularity. *Neikos* is structurally necessary not because *Philia* says so, but because the **Zero Principle** demands a background for any determinate form to exist.

A.7.2 All First Principles Grounded in GZP

- **FP1 (Primacy of the *Archē*):** The *Archē* is determinate reality, grounded by contrast with the *Apeiron*

- **FP2 (Diachronic Primacy):** Becoming is primary being; stability is pattern within flux —grounded in the contrast between process and stasis

- **FP5 (Impotence Before the *Apeiron*):** Direct application of GZP: the indeterminate cannot be known determinately

- **FP8 (Navigability):** Knowledge is possible because reality is determinate (has boundaries) against the indeterminate

A.8 The Ultimate Justification

Why accept GZP? Because:

1. **It works:** It successfully stops infinite regress and circular reasoning
2. **It's necessary:** Any system that denies it must either accept regress/circularity or posit an arbitrary stopping point
3. **It's empirically coherent:** We observe bounded things in unbounded space, specific events in undifferentiated time, meaningful concepts in semantic fields
4. **It's performatively consistent:** To deny GZP is to use determinate reasoning to argue against the necessity of determinate reasoning's ground

A.9 Conclusion: The First First-Principle

GZP is not merely another principle in the NPN system — it is **the principle that makes principles possible**. It answers the most fundamental question: "How can anything be anything?" By providing the necessary condition: **determination requires an indeterminate ground**.

All philosophy since Anaximander has been grappling with this insight, often without naming it. By making GZP explicit, NPN provides what

APPENDIX A. THE GENERAL ZERO PRINCIPLE 185

previous systems lacked: a clean, non-circular foundation that stops the infinite regress before it starts.

The *Apeiron* is not a primitive cosmological speculation. It is the necessary answer to the oldest and deepest philosophical problem: **How to ground something in nothing without making that something into just another thing.**

GZP is the **first first-principle**—not because it comes first in a list, but because it is the condition that makes any list of principles possible. By grounding determination in the indeterminate, it stops the infinite regress that has haunted philosophy since its inception. What Anaximander intuited, Aristotle formalized, and Spencer-Brown mathematized, NPN makes explicit: **reality is bounded against the boundless, and that is why it can be known at all.**

See also: Appendix C: for Anaximander's foundational insights and its etymological justification
See also: Appendix D: for Plato's application and inversion of GZP
See also: Appendix E: for Aristotle's formalization of GZP

Appendix B

The First Principles of Neo-Pre-Platonic Naturalism

All men by nature desire to know. An indication of this is the delight we take in our senses; for even apart from their usefulness they are loved for themselves... We say we know a thing only when we think we recognize its first cause [*archē*], that from which it is and cannot be otherwise. — Aristotle [1]

How fleeting are the wishes and efforts of man! how short his time! and consequently how poor will his products be, compared with those accumulated by nature during whole geological periods. Can we wonder, then, that nature's productions should be far 'truer' in character than man's productions; that they should be infinitely better adapted to the most complex conditions of life, and should plainly bear the stamp of a far higher workmanship? — Charles Darwin[2]

B.1 Introduction: The Core Principles of NPN

A philosophical system is measured by the strength of its foundations. The following principles are not merely axioms chosen for debate, but **First Principles**—fundamental, irreducible truths discovered as the necessary ground for any coherent, naturalistic account of reality, mind, and

---

   1. *Metaphysics* Book I (980a21–981b10), in *The Complete Works of Aristotle: The Revised Oxford Translation*, ed. Jonathan Barnes, vol. 2 (Princeton: Princeton University Press, 1984), p. 1552.
   2. *On the Origin of Species*, 6th ed. (London: John Murray, 1872), p. 428.

value. To deny them is not to propose an alternative system, but to collapse into absurdity or self-contradiction. They form the bedrock of the Neo-Pre-Platonic Naturalist meta-structure.

A First Principle is a foundational starting point. A **Corollary** is a proposition that follows with little or no proof from one already established. The following structure organizes these tenets according to this logical hierarchy. For the complete geometric proof of the first principle below, see Appendix K: First Philosophy: The Boundary Condition.

B.2 The Ground: The Precondition of Existence

The General Zero Principle (GZP) establishes the logical necessity of an indeterminate background for any determination whatsoever. From this general necessity, the specific ontological Zero Principle (ZP) for systems is derived.

B.2.1 GZP: GENERAL ZERO PRINCIPLE

Statement: For anything to possess determinate identity, meaning, or existence, it must exist within a delimited context set against an **indeterminate background**. The ultimate foundation cannot itself be determinate, for then it would require further foundation. It must be **indeterminate**.

Justification: Infinite regress is avoided only by positing a ground that is not a "thing" among things. To define "A" requires "not-A". If "not-A" is also determinate ("B"), the regress continues. It must stop at the indeterminate.

B.2.2 ZP: ZERO PRINCIPLE: THE NECESSITY OF CONTRAST

Statement: For any determinate system to exist, there must be an indeterminate complement — a not-system. Identity is not intrinsic but relational, defined by emergence from a contrasting field.

Justification: To define a thing is to distinguish it from what it is not. If the context of distinction were itself fully determinate, an infinite regress would render definition impossible. To say a thing has an identity is to say it emerges from a field of contrast — a ground that is not the thing.

To claim there is a system is to claim there is an "inside" to the system, which is meaningless unless there is an "outside" the system.

B.2.3 C1: PRIMARY COROLLARY: THE APEIRONIC CONTEXT

Statement: The *Apeiron* is the necessary, indeterminate field that provides the ontological contrast for the *Archē*. It is not merely an epistemic limit of knowledge, but the ontological background required for the *Archē* to possess determinate identity.

Justification: By the **Zero Principle**, identity is differential, not intrinsic. The *Archē* (the totality of determinate reality) therefore cannot be defined in isolation; it requires a complement. The *Apeiron* serves as that absolute complement—the context-field of indeterminacy that allows the *Archē* to be something rather than nothing.

B.3 The Foundation: The Nature of Reality

These principles describe the fundamental nature of the cosmos.

B.3.1 FP1: FIRST PRINCIPLE OF REALITY: THE PRIMACY OF THE *ARCHĒ*

Statement: The *Archē* is. It is the fundamental, objective, physical reality that exists. The *Archē* is the non-negotiable ground of all inquiry.

Justification: To deny it is self-refuting: the very act of denial must be performed *within* a reality by a conscious entity that *exists*. Solipsism and radical skepticism are not arguments against the *Archē*; they are linguistic games played within it.

B.3.2 FP2: FIRST PRINCIPLE OF BECOMING: DIACHRONIC PRIMACY

Statement: Being is a stabilized pattern within Becoming. A synchronic state is a derived abstraction; **Becoming** is ontologically primary.

Justification: The negation of this principle posits a world without change—a static, timeless snapshot as fundamental reality. This is not merely false but logically incoherent and a performative contradiction.

B.3.3 FP3: FIRST PRINCIPLE OF COSMIC DYNAMICS: THE LOGOS AND ITS EXHAUSTIVE POLARITY

Statement: The constitutive interaction of the *Archē* is grounded in the Zero Principle of Contrast. The *Logos*—the potential for all relation—is exhaustively and necessarily actualized through the fundamental polarity of attraction (*Philia*) and repulsion (*Neikos*). These forces are the mutually constitutive context-fields for one another; a third fundamental mode of relation is metaphysically and conceptually impossible.

Justification: This follows directly from the Zero Principle: for any system to possess identity, it must emerge from a field of contrast.

1. **Relational Necessity:** *Philia* (Union) is only intelligible against the background of *Neikos* (Division). To "unite" is to overcome a state of separation; to "separate" is to break a state of union. Neither force can exist in isolation, as each provides the "outside" required for the other's "inside."
2. **Logical Exhaustion:** Any conceivable interaction within the *Archē* must either move entities "toward" (Union) or "away from" (Division) one another. To posit a third mode that is neither unifying nor dividing is to posit a vector without direction—a logical nullity.
3. **Ontological Closure:** Because identity is relational (ZP), the *Logos* must manifest as a polarity. A singular force would lack a contrasting background, collapsing the *Archē* into an undifferentiated, static unity where change—and thus the *Archē* itself—would cease to exist. Conversely, absolute division without the unifying force of *Philia* would prevent the formation of any persisting systems, leading to a total fragmentation into the *Apeiron* where identity is likewise impossible.

APPENDIX B. FIRST PRINCIPLES OF NPN

B.3.4 FP4: FIRST PRINCIPLE OF EMERGENT COMPLEXITY: THE POTENTIAL FOR *NOUS*

Statement: The constitutive interaction of *Philia* and *Neikos* within the *Logos* is inherently generative, making the emergence of complex, stratified systems—including the *Nous*—a natural and potential outcome of the *Archē*'s dynamics, not a supernatural accident.

Justification: The negation leads to a double contradiction.

1. **Ontological contradiction:** Since the *Archē* is the sole fundamental reality (FP1), any phenomenon that exists must have its sufficient cause within the *Archē*. The *Nous* exists; therefore, the generative potential for the *Nous* must be a constitutive property of the *Archē*.

2. **Performative contradiction:** To claim that the *Archē* cannot produce the *Nous* is to use the *Nous* to deny its own origin. It is the act of a product denying the capacity of its own factory. To argue the point is to prove it false.

B.4 The Boundary: The Limits of Knowledge

These principles define the scope and nature of what can be known.

B.4.1 FP5: FIRST PRINCIPLE OF KNOWLEDGE: THE IMPOTENCE BEFORE THE *APEIRON*

Statement: The *Apeiron* is the category for which empirical observation is impossible. Therefore, no logical operation can be empirically grounded or validated.

Justification: Logic is the *Somatic logos*—a tool for processing empirical observations. To apply logic to a domain with zero observations is to run a program with no input data. The output is not knowledge, but nonsense. The negation is an attempt to generate information from an information vacuum, which is a logical absurdity.

B.4.2 C2: PRIMARY COROLLARY: THE CONFIDENCE GRADIENT OF EPISTEMIC JUSTIFICATION

Statement: *Epistēmē* is a provisional, high-fidelity model of the *Logos*, justified by its predictive success and functional utility in navigating the *Archē*. Its measure is a **Confidence Gradient**, not a binary state of certainty.

Justification: This is the direct epistemological consequence of the **First Principle of Diachronic Primacy**. A finite model generated *within* a system cannot contain a perfect, complete representation of that system. The negation—demanding certain, final knowledge—is a logical impossibility, as it requires that a finite *Nous*, itself a small subset of the *Archē*, could generate a synchronic snapshot of the *Archē* that is certain, complete, and final. The Confidence Gradient is therefore the *only possible form* that justified knowledge can take for a finite, emergent navigator.

B.5 The Bridge: The Principles of Relation

These principles explain how a knower can reliably interact with and understand reality.

B.5.1 FP6: FIRST PRINCIPLE OF AGENCY: THE PRIMACY OF THE *HORMĒ*

Statement: The *Hormē* is the constitutive, non-negotiable ground of being an agent.

Justification: The negation is a functional absurdity that violates the causal structure of reality. To assert that an agent exists without *Hormē* is to claim that an inert system can persist against the entropic forces of *Neikos*. In a dynamic *Archē*, any system that fails to strive is strictly identical to a decaying object. Therefore, it is a performative contradiction to assert that an agent—an entity defined by its persistence—can exist without the striving necessary to endure through deep time.

B.5.2 FP7: FIRST PRINCIPLE OF EPISTEMIC GROUNDING: THE *SOMATIC LOGOS*

Statement: The operational logic of the evolved *Nous* is functionally aligned with the operational *Logos* of the *Archē* because the former is a product of the latter.

Justification: The negation is a performative contradiction. A *Nous* fundamentally misaligned with the *Logos* would produce maladaptive models, leading to extinction long before it could develop the capacity for abstract inquiry. To deny this principle is to use a functioning navigator to argue for the impossibility of navigation.

B.5.3 C3: COROLLARY: THE SOMATIC PRESENT AS A NAVIGATIONAL SNAPSHOT

Statement: The conscious experience of a synchronic "now" is a computational necessity for any finite Navigator. It is the operational interface that collapses the diachronic complexity of the *Archē* into an actionable moment.

Justification: This follows directly from the First Principle of Becoming: Diachronic Primacy. Because the *Archē* is a continuous flux, it presents an infinite stream of causal data. A finite Navigator cannot process infinite continuity in real-time. Therefore, to calculate a decision, the *Nous* must perform a **Dimensional Reduction**—freezing the 4D process into a static 3D model (the snapshot). The negation is a mathematical absurdity: to claim a finite agent can act without a snapshot is to assert it can calculate a vector of action without a coordinate of origin. Since a dynamic flux offers no inherent fixed points, the snapshot is the non-negotiable zero-point required for the computation of agency.

B.6 The Navigator: The Emergent Agent

These principles define the nature of the knower and the foundation of value.

B.6.1 FP8: FIRST PRINCIPLE OF RELATION: THE NAVIGABILITY OF THE *ARCHĒ*

Statement: The *Logos* of the *Archē* is, in principle, model-able by an emergent subsystem within it. Reality is structured such that it can be successfully navigated.

Justification: The negation is a performative contradiction. To claim "reality is fundamentally unnavigable" is to assert a model of reality (the "unnavigable" model) that you are, by the act of asserting it, successfully using to navigate. It is the claim, "My map of un-mappability is correct." This is the deepest possible self-refutation.

B.6.2 FP9: FIRST PRINCIPLE OF THE *NOUS*: META-COGNITIVE POTENTIAL

Statement: The *Nous* is inherently capable of reflexive self-modeling. It can form models not only of the external *Archē* but also of its own processes, states, and models. This self-referential capacity is not merely an observed property but a *transcendental condition* for any system that can formulate propositions about its own nature.

Justification: The negation is a performative contradiction. To claim "the *Nous* cannot model itself" is to *present a model of the Nous*. The act of denial is itself an instance of the capacity being denied.

B.6.3 C4: PRIMARY COROLLARY: THE OBJECTIVITY OF VALUE

Statement: For any system possessing a constitutive *Hormē*, 'good' is that which fulfills its striving and 'bad' is that which frustrates it. Value is an objective, functional relationship between a system's states/actions and the successful expression of its *Hormē* within the constraints of the *Logos*. The *Hormē* is the non-negotiable ground of all teleology, and thus of all value.

Justification: This is the direct ethical consequence of the **First Principle of Agency: The Primacy of the *Hormē***. A system that strives *is* a system for which states of the world can be functionally categorized as beneficial or detrimental to its core operational imperative. To deny this is to assert there is no difference between a flourishing and a dying system, which is empirically false and logically absurd.

B.7 Theorems: Derived Doctrines

These are major conclusions and methodologies derived from the interaction of the First Principles and their Corollaries. They represent the applied power of the NPN meta-structure.

B.7.1 T1: THEOREM: THE POPPERIAN PROTOCOL

Statement: The complete methodology of conjecture and refutation (falsification) is the necessary and optimal procedure for a finite *Navigator* to increase the Confidence Gradient of its models of the *Archē*. It is the systematic implementation of the epistemic cycle of *Lysis* and *Aporia*.

Derivation: This protocol is logically entailed by the First Principles of Neo-Pre-Platonic Naturalism.

1. The *Archē* exists as an objective reality (Primacy of the *Archē*), and a finite *Navigator* has emerged within it (Potential for *Nous*). The *Navigator*'s constitutive task is therefore to build functional models of this reality.

2. However, the *Archē* is fundamentally a process, not a static state (Diachronic Primacy). Consequently, no model can be a certain or final snapshot of reality.

3. It follows that knowledge must be a Confidence Gradient—a measure of a model's predictive success and utility—rather than a state of certain possession (Corollary: The Confidence Gradient).

4. Critically, the *Archē* is bounded by the *Apeiron*, the permanently unknowable (Impotence Before the *Apeiron*). This guarantees a permanent frontier of the "not-yet-known *Archē*" that can potentially contradict any current model.

5. Therefore, the only way to increase a model's Confidence Gradient is to test it aggressively against this frontier. A test that a model fails induces a cognitive *Lysis* and productive *Aporia*, clearing the ground for a better model. A test a model survives raises its justified confidence. This process of testing—of attempted refutation—*is* the logic of falsification.

Conclusion: Falsification is not merely a useful heuristic. It is the necessary and defining activity of a finite *Navigator* within a diachronic, boundless *Archē*. It is the engine of the Socratic *Elenchus* applied to the cosmos, and the only non-arbitrary procedure for the growth of knowledge.

B.7.2 T2: THEOREM: THE STATUS OF FORMAL TRUTHS

Statement: Analytic truths (logic, mathematics) are certain within their defined contrast-domains because their boundaries are stipulated. Synthetic knowledge — modeling the *Arche* — is gradient-bound because its boundary is the *Apeiron*, which cannot be stipulated or observed.

Derivation:

1. By **ZP**, identity requires contrast.
2. In analytic systems, contrast is stipulated and controlled → **certainty**.
3. The *Archē* is contrasted by the *Apeiron* (**FP4**) — uncontrolled and un-model-able.
4. Therefore, synthetic knowledge is gradient-bound (**C2**).
5. Applying analytic systems to the *Archē* treats them as synthetic models → subject to **C2**.

Conclusion: Analytic truth is grammar-certain. Synthetic truth is map-provisional. The Navigator uses the first to refine the second, always acknowledging the boundary between defined contrast and given contrast — the *Apeiron*.

B.7.3 T3: THEOREM: THE NECESSARY DISTORTION

Statement: Every synchronic model generated by the Nous is strictly ontologically distinct from and descriptively incomplete relative to the diachronic Archē it represents. Therefore, "distortion" is not a failure of cognition, but a structural requirement for navigation.

Derivation: This theorem describes the inevitable gap between the map and the territory.

1. The Archē is a continuous, unfolding process of Becoming (**FP2: Diachronic Primacy**).
2. The Navigator must perform a dimensional reduction, freezing the 4D process into a static 3D snapshot to make a decision (**C3: The Somatic Present**).
3. A static snapshot (t_0) cannot contain the vector of change ($t_0 \to t_1$) without converting it into a static symbol.
4. Therefore, the model must fundamentally differ in nature from the reality it models to be usable.

Conclusion: Just as a photograph is a distortion of movement, a concept is a distortion of process. The Navigator must trade *fidelity* (completeness) for *utility* (actionability).

B.7.4 T4: THEOREM: ETHICAL ISOMORPHISM

Statement: Epistemic error (falsehood) and Ethical vice (immorality) are functionally isomorphic; both are states of misalignment between the Navigator's internal models and the external Logos. Therefore, the pursuit of Truth and the pursuit of the Good are the same mechanical operation.

Derivation: This theorem unifies the epistemic and ethical branches of the system.

1. **Definition of Good:** "Good" is defined as the functional alignment of a system's actions with the Logos to fulfill its Hormē (**C4: The Objectivity of Value**).

2. **Definition of Truth:** "Truth" (or valid reasoning) is the functional alignment of the Nous's logic with the causal structure of the Archē (**FP7: The Somatic Logos**).
3. Since both Good and Truth are defined by the single metric of *Functional Alignment with the Logos*, a failure in one is necessarily a failure in the other.
4. Maladaptive behavior (vice) is the result of acting on a map that does not match the territory.

Conclusion: An unethical act is an act based on a false model of reality. To improve character is to improve one's map.

B.7.5 T5: THEOREM: THE ENTROPIC MANDATE

Statement: Any system that attempts to permanently suppress Neikos (dissent, variation, friction) in favor of pure Philia (unity, stability) guarantees its own entropic collapse. Stability is not a static state, but a dynamic oscillation.

Derivation: This theorem applies cosmic dynamics to political and systems theory.

1. A system strives to persist (**FP6: Primacy of the Hormē**), but the environment (Archē) is constantly changing (**FP2: Diachronic Primacy**).
2. Persistence therefore requires adaptation.
3. Adaptation requires Neikos (differentiation/selection) to break down maladaptive structures and generate new options (**FP3: Logos and Polarity**).
4. A system that eliminates Neikos eliminates its ability to adapt to the changing Archē.

Conclusion: "Perfect order" (Pure Philia) is functionally identical to death. A healthy, persisting system *must* institutionalize controlled *Neikos*.

Here is **Theorem T5** formatted in Markdown to match the visual structure of your previous theorems, followed by the LaTeX code for your book layout.

B.7.6 T6: THE LIFE-AGENCY ISOMORPHISM THEOREM

Statement: Life and minimal agency are isomorphic. A system is alive if and only if it possesses *Hormē* (the striving to persist), and it possesses *Hormē* if and only if it is an agent. The capacity to strive—to maintain far-from-equilibrium organization against entropy—is the constitutive property of both biological existence and navigational agency.

Derivation:

1. By thermodynamic necessity, living systems must perform continuous work to maintain low-entropy states against environmental dissolution.
2. This work is functionally identical to *Hormē*—the striving to persist (see FP6).
3. Agency is defined as the capacity to act in ways that bias outcomes toward persistence.
4. Therefore, the metabolic work that defines life **is** the striving that defines agency.
5. The isomorphism is scale-invariant: from bacterial chemotaxis to human deliberation, all are expressions of the same anti-entropic drive.

Implications:

- **Dissolution of the Gap:** It bridges "Is" and "Ought." The "Fact" of life is inherently directional; to *be* is to *ought to persist*.
- **The Ontological Boundary:** It distinguishes Agents from Objects. Passive systems (rocks, tools, machines) cannot be agents because their persistence is not thermodynamically linked to their performance. They function, but they do not strive.

- **Evolutionary Continuity:** Human ethics is not a magical addition to matter but the sophisticated elaboration of bacterial thermodynamics.

B.7.7 T7: THEOREM: THE ENTROPIC ASYMMETRY (THE COST OF BEING)

Statement: The maintenance of any intelligible pattern (*Being*) within the indeterminate *Apeiron* requires the continuous expenditure of energy (*Hormē*) against a standing gradient of dissolution. Order is statistically unlikely and energetically expensive; disorder is the statistical default and energetically free.

Derivation:

1. By **FP2**, *Becoming* (flux) is ontologically primary; stable *Being* is a temporary restriction of potentiality.
2. By **GZP**, the ground is the indeterminate *Apeiron*.
3. Any determinate pattern requires a boundary that excludes noise; this exclusion creates a thermodynamic gradient.
4. Therefore, to persist is to actively resist the statistical probability of returning to the *Apeiron* (dissolution).
5. **FP6** (*Hormē*) is the constitutive force required to pay this energetic cost.

Conclusion: Existence is resistance. Order is not a natural resting place, but a wage continuously earned against the void. The Navigator understands that "Good" is the efficient generation of *negentropy* (coherent order), while "Evil" is often merely the cessation of the work required to maintain the boundary.

APPENDIX B. FIRST PRINCIPLES OF NPN

Table B.1 Summary of First Principles, Corollaries, and Theorems

Principle / Doctrine	Core Statement
GZP: General Zero Principle	All determination requires an indeterminate ground. *(Logical Necessity)*
ZP: Contrast and Precondition	For a system to exist, it must possess an "outside" or contrasting background from which it is distinguished.
FP1: Primacy of the *Archē*	The *Archē* is the fundamental, objective, physical reality that exists.
FP2: Diachronic Primacy	Being is a stabilized pattern within Becoming. *Becoming is ontologically primary.*
FP3: The *Logos* and Polarity	The potential for relation is actualized through attraction (*Philia*) and repulsion (*Neikos*).
FP4: The Potential for *Nous*	The emergence of *Nous* is a natural outcome of the *Archē*'s dynamics.
FP5: Impotence Before the *Apeiron*	Logical operation cannot be empirically grounded within the unknowable *Apeiron*.
FP6: Primacy of the *Hormē*	The *Hormē* is the constitutive, non-negotiable ground of being an agent.
FP7: The *Somatic logos*	The logic of the *Nous* is an evolved functional alignment with the *Logos* of the *Archē*.
FP8: Navigability of the *Archē*	The *Logos* of the *Archē* is, in principle, model-able by an emergent subsystem.
FP9: Meta-Cognitive Potential	The *Nous* is inherently capable of reflexive self-modeling.
C1: The Apeironic Context	The *Apeiron* is the necessary, indeterminate field that provides the ontological contrast for the *Archē*.
C2: The Confidence Gradient	*Epistēmē* is a provisional model measured on a gradient, not a binary of certainty.
C3: The Somatic Present	The conscious "now" is a necessary, evolved heuristic of the *Somatic logos*.
C4: The Objectivity of Value	"Good" is that which fulfills a system's *Hormē*; "bad" is that which frustrates it.

Principle / Doctrine	Core Statement
T1: The Popperian Protocol	Falsification is the necessary procedure to increase the Confidence Gradient of models.
T2: The Status of Formal Truths	Analytic truths retain formal certainty but must be evaluated on the gradient when applied to the *Archē*.
T3: The Necessary Distortion	Synchronic models are descriptively incomplete relative to diachronic reality; *distortion is required for navigation.*
T4: Ethical Isomorphism	Epistemic error and ethical vice are functionally identical states of misalignment with the *Logos*.
T5: The Entropic Mandate	Any system that suppresses *Neikos* (dissent/variation) guarantees its own entropic collapse.
T6: Life-Agency Isomorphism	Life and minimal agency are isomorphic; the thermodynamic work of persistence (*Hormē*) is the constitutive property of both biological existence and navigational agency.
T7: The Entropic Asymmetry	The maintenance of any intelligible pattern (*Being*) within the indeterminate *Apeiron* requires the continuous expenditure of energy (*Hormē*) against a standing gradient of dissolution.

Appendix C

Anaximander and the Zero Principle: The Life-Cycle of Identity

"Determination is negation." — Benedict de Spinoza[1]

C.1 The First Philosopher of Relation

Anaximander of Miletus (c. 610–546 BCE) did not ask *what* things are made of. He asked something far more profound: **What does it mean to be something rather than nothing?** His answer—preserved in a single, dense fragment reported by Simplicius—describes not the substance of reality, but **the life-cycle of identity itself.**[2]

C.1.1 The Etymological Foundation

The key to his system is relational from the ground up, as revealed in the very structure of his terms:

- *A-peiron* (Ἄπειρον) — **not-bounded, not-limited**. From *a-* (privative prefix "not") + *peirar* ("boundary, limit").[3]

---

1. Benedict de Spinoza, *Complete Works*, ed. Michael L. Morgan, trans. Samuel Shirley (Indianapolis: Hackett Publishing, 2002), 892.

2. The fragment is cataloged as DK 12 B 1 in the standard collection: Hermann Diels and Walther Kranz, *Die Fragmente der Vorsokratiker*, 6th ed. (Berlin: Weidmann, 1951). The text is preserved in Simplicius, *Commentary on Aristotle's Physics*, 24.13.

3. Pierre Chantraine, *Dictionnaire étymologique de la langue grecque* (Paris: Klincksieck, 1968), s.v. "πέρας." See also Robert Beekes, *Etymological Dictionary of Greek* (Leiden: Brill, 2010), s.v. "ἄπειρος."

- ***A-dikia*** (Ἀδικία) — **not-just, injustice**. From *a-* ("not") + *dikē* ("justice, judgment"). The root *dik-* relates to "pointing out," "showing," "separating"—a judgment that distinguishes right from wrong. This is the act of being seperated from the *Apeiron*.[4]

- ***Di-kē*** (Δίκη) — **the cut, the division, the measure**. The term carries connotations of judgment as a **separating action**—cutting truth from falsehood, right from wrong. In Anaximander's context, it becomes an ontological principle of re-integration into the *Apeiron*.[5]

These are not substances. They are **relations**. The very morphology of Anaximander's vocabulary reveals a thinker working in terms of **contrast, separation, and boundary-maintenance**—not in terms of material constituents.[6]

C.1.2 The Four-Step Cycle of Identity

Anaximander's fragment outlines a complete metaphysical process in four steps:[7]

---

4. Beekes, *Etymological Dictionary*, s.v. "δίκη." Compare Chantraine, *Dictionnaire*, s.v. "δίκη," noting its connection to the notion of "showing" or "indicating."

5. For the judicial and cosmic aspects of *dikē* in early Greek thought, see Hugh Lloyd-Jones, *The Justice of Zeus*, 2nd ed. (Berkeley: University of California Press, 1983), 32–35; and Gregory Vlastos, "Equality and Justice in Early Greek Cosmologies," *Classical Philology* 42 (1947): 156–78. See also Charles H. Kahn, *Anaximander and the Origins of Greek Cosmology* (New York: Columbia University Press, 1960), 178–83, who argues that *dikē* here represents the mutual reparation of opposites.

6. For recent reassessments of Anaximander's systematic and proto-scientific thinking, see Daniel W. Graham, *Explaining the Cosmos: The Ionian Tradition of Scientific Philosophy* (Princeton: Princeton University Press, 2006), 29–56; and Andrew Gregory, *Anaximander: A Re-assessment* (London: Bloomsbury Academic, 2016), 43–83.

7. This four-step reconstruction, while not explicitly advanced in the secondary literature, draws systematically on the etymological and philosophical analysis of Kahn, *Anaximander*, 166–98; G. S. Kirk, J. E. Raven, and M. Schofield, *The Presocratic Philosophers: A Critical History with a Selection of Texts*, 2nd ed. (Cambridge: Cambridge University Press, 1983), 117–21; and the relational ontology developed in this volume.

Step 1: The Boundless Ground (*Apeiron*)

The *Apeiron* is the **indeterminate, unbounded whole**—the necessary background against which anything determinate can appear. It is not "stuff," but the **relational precondition** for boundedness. In NPN terms, this is the **indeterminate complement required by the Zero Principle**.[8]

Step 2: The Cut of Individuation (*Adikia*)

To become a thing is to be **cut out** from the *Apeiron*. This act of separation is the primal "injustice" (*adikia*)—the violation of primordial unity. Identity is born through **contrast**: being something means **not being everything else**.[9]

Step 3: The Measure of Distinction (Time)

Once cut out, a form persists. **Time measures how long this distinction can be maintained** against the dissolving pull of the boundless. Time is not a passive dimension but an **active assessment of contrast-persistence**.[10]

Step 4: The Re-Cut of Dissolution (*Dikē*)

No distinction is permanent. The cosmic principle of *Dikē*—often mistranslated as "justice"—is the **force that cuts the form back into the Apeiron**. This is the "penalty" for the "injustice" of individuation: the necessary re-integration of the temporary into the eternal.

---

8. Kahn, *Anaximander*, 238: "The Unlimited is not a qualitatively distinct element... but the inexhaustible source from which all particular forms emerge."

9. See Kirk, Raven, and Schofield, *The Presocratic Philosophers*, 119, discussing the "unjust" encroachment of opposites upon one another.

10. The Greek phrase is *kata tēn tou chronou taxin* ("according to the ordering of Time"). Kahn interprets this as the "ordinance" or "assessment" that limits the duration of any finite existence (Kahn, *Anaximander*, 172). Kirk, Raven, and Schofield render it "according to the assessment of Time" (Kirk, Raven, and Schofield, *The Presocratic Philosophers*, 118). Jonathan Barnes translates it "as is ordained by Time" (Barnes, *Early Greek Philosophy*, 2nd ed. (London: Penguin, 2001), 29). All agree Time plays an active, normative role in limiting existence.

Anaximander's cycle is a precise expression of the Zero Principle:

For any determinate system to exist, there must be an indeterminate complement.

Table C.1   Anaximander's Four-Step Cycle in NPN Terms

Anaximander's Step	NPN Formalization
1. The Boundless *Apeiron*	**Indeterminate complement** required by ZP
2. The Cut *Adikia*	**Emergence of determinate system** through contrast
3. The Measure (Time)	**Duration of contrast-maintenance**
4. The Re-Cut *Dikē*	**Dissolution of contrast**—return to indeterminate complement

C.1.3   Time as the Assessor of Distinction

Anaximander's phrase "according to the assessment of Time" (*kata tēn tou chronou taxin*) is revolutionary. He sees Time not as a container but as an **active measure of how long a form can hold its boundaries.** This anticipates:

- **FP2: Diachronic Primacy** — Being is pattern within Becoming.
- **T5: The Entropic Mandate** — All forms eventually dissolve.
- **The Confidence Gradient** — All knowledge is provisional across time.

Time is the universe's way of **measuring the persistence of contrast.**

C.1.4   Note on the Mechanics of Adikia and Dikē

If we view the cycle through the lens of the exhaustive polarity, the role of the forces becomes clear:

- **Adikia (Injustice) is Sustained Philia**: To exist as a determinate form is to maintain a localized cohesion. The "injustice" is the act

of *Philia* holding elements together in a specific configuration, preventing them from returning to the boundless potential of the *Apeiron*. It is a temporary monopoly on existence.
- **Dikē (Justice) is the Work of Neikos**: The "penalty" paid to Time is the eventual dissolution of this cohesion. *Neikos* breaks the internal bonds of the form, separating the parts so they may be reintegrated into the undifferentiated whole. In this sense, entropy is the executioner of cosmic justice.

C.2 Why Aristotle Could Not See It

Aristotle's metaphysics was built on **substance** (*ousia*) as primary reality. When he read Anaximander, he translated the relational terms into substance-categories:

Table C.2   Anaximander's Relational Ontology

Anaximander's Relational Ontology	Aristotle's Substance Reading
Apeiron as contrast-field	*Apeiron* as indefinite matter
Adikia as state of separation	*Adikia* as moral wrong
Dikē as re-integration force	*Dikē* as legal justice
Identity as temporary contrast	**Identity as essential nature**

This misreading was not an isolated error. Aristotle's entire philosophical system was grounded in the primacy of substance. Consequently, he systematically reinterpreted the Pre-Platonics—who were largely concerned with processes, relations, and dynamic principles—through the lens of static substances and their attributes.[11]

---

11. See Aristotle, *Physics* III.4, 203b7–15, in *The Complete Works of Aristotle: The Revised Oxford Translation*, ed. Jonathan Barnes, vol. 1 (Princeton: Princeton University Press, 1984), 355; Aristotle, *Metaphysics* XII.2, 1069b20–22, in *Complete Works*, 2:1693.

For Aristotle, the fundamental question was: *What is the underlying substance of which all things are composed?* This led him to read Anaximander's *Apeiron* as a kind of primordial stuff—albeit indeterminate—rather than as the necessary contrast-field required for identity.[12]

The result was a tradition that, until the rise of process philosophy in the 20th century, largely overlooked the relational, dynamic, and contrast-based ontology that Anaximander and his contemporaries had pioneered. By recovering the original relational insights, NPN seeks to undo this centuries-long misdirection and return to the first principles of philosophy.

C.3   Conclusion: Recovering the First Metaphysics

Anaximander was not a primitive materialist. He was the **first philosopher of relation**, describing the life-cycle of identity with astonishing precision:

1. **Identity requires contrast** (ZP)
2. **Contrast is temporary** (enforced by *Dikē*)
3. **Time measures this temporariness**

His system is a **relational ontology** that pre-dates the substance-turn of Aristotle by two centuries. It is a metaphysics of **process, not permanence**; of **relation, not substance**; of **becoming, not being**.

The Zero Principle is not a new discovery. It is the **formalization of Anaximander's insight**—the insight that to be something is to be temporarily cut out from the boundless, measured by time, and eventually cut back in[13].

---

12. Werner Jaeger famously argued that Aristotle projected his own concept of *hylē* (matter) back onto the Presocratics, distorting their original metaphysical insights. See Werner Jaeger, *The Theology of the Early Greek Philosophers* (Oxford: Clarendon Press, 1947), 24–37.

13. This reading aligns with a growing "relational turn" in Presocratic scholarship that seeks to recover the dynamic, process-oriented metaphysics obscured by Aristotle's substance-based framework. See, e.g., John Palmer, *Parmenides and Presocratic Philosophy* (Oxford: Oxford University Press, 2009), 1–15; and Patricia Curd and Daniel W. Graham,

APPENDIX C. ANAXIMANDER AND THE ZERO PRINCIPLE

In recovering Anaximander, we recover the origin of Western philosophy as a search for **the logic of identity itself**. And in formalizing his insight as ZP, we give that search a rigorous foundation for the first time.

---

eds., *The Oxford Handbook of Presocratic Philosophy* (Oxford: Oxford University Press, 2008).

Appendix D

Plato and the Zero Principle: The Determinate *Apeiron*

"It is by beauty that beautiful things are beautiful." — Plato[1]

D.1 Introduction: Plato's Unrecognized Foundation

Plato's Theory of Forms is typically interpreted as an idealist metaphysics or an epistemological doctrine about universals. However, a Neo-Pre-Platonic Naturalist reading reveals a deeper structure: Plato's system is a monumental, if inverted, application of what was formalized as the **General Zero Principle** in Appendix A. He recognized that determinate particulars require a contrasting ground, but he characterized that ground as *more* determinate rather than *less*—creating a system that is structurally sound but metaphysically inverted. This appendix traces how Plato intuitively grasped **GZP**, built his philosophy upon it, and spent his late career confronting the contradictions that arose from his inversion.

D.2 The Middle Dialogues: Building the Determinate Background

The heart of Plato's philosophical project in his middle period—expressed most famously in the *Phaedo*, *Republic*, and *Symposium*—is the Theory of Forms. While traditionally read as a doctrine of transcendent idealism or a theory of universals, a Neo-Pre-Platonic Naturalist analysis reveals its

---

1. Plato, *Phaedo* 100c. This tautology is the structural flaw NPN identifies: using a determinate concept (Beauty itself) to ground determinate instances, rather than an indeterminate ground. In *Plato: Complete Works*, ed. John M. Cooper (Indianapolis, IN: Hackett Publishing Company, 1997).

deeper structural logic: it is a monumental, if inverted, application of the Zero Principle.

D.2.1 The Theory of Forms as Applied Zero Principle

In dialogues such as the *Phaedo*, *Republic*, and *Symposium*, Plato presents the Forms as eternal, perfect, and unchanging realities. Particulars in the sensible world are imperfect copies or participants in these Forms. Structurally, this is a clear application of the Zero Principle:

- **Determinate System:** The world of changing, imperfect particulars.
- **Indeterminate Complement Required:** A contrasting background that explains their nature and provides their identity.
- **Plato's Solution:** The Realm of Forms—a *determinate* background of perfect paradigms.

Where Anaximander correctly posited an *indeterminate* background (*Apeiron*), Plato supplied a *hyper-determinate* one. The logic is ZP-compliant: particulars need a "not-this" to be "this." But the content is inverted: the "not-this" becomes a realm of perfect "thats."

D.2.2 Why Plato's Inversion Is Compelling

Plato's move is psychologically and logically natural. The human mind struggles to conceive of pure indeterminacy. When asked, "What is the background against which particulars exist?" it is easier to imagine a *better, more real* version of those particulars than to conceive of a formless void. The Forms are the *Apeiron* made thinkable—the indeterminate rendered as a superlative determinate.

This inversion is not arbitrary. In a profound sense, the Realm of Forms represents a **positive definition of indeterminacy**: it is the *totality of all perfect possibilities*. To be "all perfect forms" is to be no particular form—it is a *determinate indeterminacy*. Where Anaximander's *Apeiron* is the *negative indeterminate* (the boundless, the unlimited, the not-this), Plato's Forms are the *positive indeterminate* (the all-perfect, the complete

APPENDIX D. PLATO AND THE ZERO PRINCIPLE 213

set, the everything-of-its-kind). Both serve the same Zero Principle function: they provide the relational contrast that gives any particular thing its identity and meaning. The Form of the Good, as the ultimate, contentless source of all Forms, functions almost exactly as the *Apeiron*—it is the indeterminate ground of all determination.

This inversion gives his system immense explanatory power and intuitive appeal, but at the cost of metaphysical coherence. By making the contrast-field determinate, Plato substitutes a **map of perfection** for the **territory of groundlessness**—a swap that would eventually force him to confront the logical consequences in his later work.

D.3 The Late Dialogues: The System Under Stress

Having built a metaphysical edifice on the inversion of the Zero Principle, Plato's later thought can be read as a prolonged, often implicit confrontation with its structural flaws. The confident system-builder of the middle period gives way to a more tentative, self-critical philosopher whose dialogues dismantle the very foundations he once laid. This is not a retreat from philosophy, but philosophy doing its most necessary work: turning its tools upon itself.

D.3.1 The *Parmenides*: The Regress Exposed

In the *Parmenides*, Plato turns his dialectical method against his own theory. The famous "Third Man Argument" demonstrates that if a Form exists to explain the similarity of a set of particulars, then a higher Form is required to explain the similarity between the first Form and the particulars, leading to infinite regress.[2]

This is not merely a logical puzzle; it is the **General Zero Principle enforcing itself**. If the background (Forms) is itself determinate, it too requires a background. Plato's dialogue shows the infinite regress that

---

2. Plato, *Parmenides*, 132a–133a. NPN interprets this as Plato's own admission that a determinate ground (Forms) cannot solve the regress problem. A determinate Form is just another "thing" requiring explanation.

results from failing to recognize that the ultimate ground must be *indeterminate*.

NPN Interpretation: The "Third Man" regress is the inevitable outcome of trying to fulfill the Zero Principle with a *determinate* contrast-field. The Forms, as perfect paradigms, are still determinate "things." By ZP, they too need an outside. Plato's dialogue is his own system revealing that a determinate *Apeiron* is a contradiction in terms. The argument doesn't just criticize "participation"; it demonstrates that **no chain of determinate grounding can ever be complete**. The only exit is to acknowledge an indeterminate terminus—the true *Apeiron*.

D.3.2 The *Sophist*: Rehabilitating Non-Being

In the *Sophist*, Plato (through the Eleatic Stranger) makes a crucial advance: he rehabilitates "Non-Being" not as absolute nothingness, but as **Difference** (*thateron*).[3] "What is not" is simply "what is other than." This move reintroduces relationality into ontology. Being is now understood through a network of sameness and difference—a structured, differential system that mirrors the interplay of *Philia* (sameness) and *Neikos* (difference) within the *Logos*.

Here, Plato edges toward a corrective: if identity is relational and differential, then the background need not be a separate realm of perfect beings, but the **field of contrast within the One Being itself**. This anticipates the NPN view that the *Archē* contains its own exhaustive polarity.

D.3.3 The *Timaeus*:
The "Likely Story" and the Retreat from Certainty

The *Timaeus* presents a cosmological myth where a divine craftsman (*Demiourgos*) looks to the Forms and imposes order on a pre-existing

---

3. Plato, *Sophist*, 257b–259b. This marks the transition from *Static Being* to *Relational Ontology*, anticipating the NPN concept of differentiation (*Neikos*) as a constitutive force.

APPENDIX D. PLATO AND THE ZERO PRINCIPLE

chaotic receptacle (*khōra*). Crucially, Timaeus repeatedly calls his account a "likely story" (*eikos mythos*).[4]

This is often read as epistemological humility, but from a GZP perspective, it is more profound: it is the **acknowledgment that the ultimate ground cannot be captured in a determinate account**. The *khōra*—the receptive, formless space—functions as an inscrutable, quasi-indeterminate matrix. Plato is no longer trying to define the background; he is gesturing toward its necessary indeterminacy.

D.4 Plato's Struggle Towards General Zero Principle

Plato's philosophical development can be read as a prolonged, indirect confrontation with the demands of the General Zero Principle:

1. **Middle Period:** Applies ZP by creating a determinate background (Forms) for determinate particulars.
2. ***Parmenides***: Exposes the infinite regress that results from a determinate background.
3. ***Sophist***: Begins to correct by making Being inherently relational (differential).
4. ***Timaeus***: Retreats to a "likely story," acknowledging the indeterminacy of the ground (*khōra*).

He never names the *Apeiron*, but his late works strain toward its conceptual space. The Forms, as a determinate *Apeiron*, could not hold.

D.5 Why Plato Matters for NPN

Plato's value to Neo-Pre-Platonic Naturalism is twofold:

1. **As a Negative Example:** His inversion shows the peril of making the contrast-field determinate. The infinite regress of the "Third Man" is the logical price of ignoring GZP's demand for an indeterminate ground.

---

4. Plato, *Timaeus* 29b–d. NPN reads this not as a lack of rigor, but as the first formal acknowledgment of the *Confidence Gradient*: knowledge of ultimate origins is inherently probabilistic.

2. **As a Methodological Pioneer:** His dialectic, his exploration of non-being as difference, and his late epistemological humility provide tools and insights that remain valuable within a correct, GZP-grounded framework.

Plato did not have the Zero Principle, but his entire system is a **gigantic experiment in what happens when you try to build a metaphysics that respects the need for a contrast-field without accepting its indeterminacy**. The result is both brilliant and broken—a castle in the sky that he himself spent his final years dismantling.

D.6 Conclusion: Plato's Tragic Genius

Plato saw the problem that Anaximander saw: determinate things require a ground. He offered the most intellectually compelling wrong answer in the history of philosophy: a ground made of perfected versions of those things. His late dialogues record his dawning realization that this answer would not hold. In struggling with the regress, rehabilitating non-being, and retreating to "likely stories," he stumbled toward the very principle he could not quite name.

The Theory of Forms is not a naïve idealism to be discarded. It is a **profound misapplication of a profound truth**—the truth that would later be formalized as the **General Zero Principle**. To study Plato through the lens of **GZP** is not to diminish him, but to understand the depth of the problem he faced and the tragic grandeur of his incomplete solution. He built the most magnificent castle in the sky because he understood, better than anyone before him, that we cannot live on the ground alone.

See also: Appendix A for the full derivation of the General Zero Principle
See also: Appendix C for the specific ontological case (ZP)

Appendix E

Aristotle and the Zero Principle: Potentiality as the Formalized Ground

> "The infinite is either that which is incapable of being traversed because it is not its nature to be traversed... or that which admits only of incomplete traversal... or that which scarcely admits of traversal." — Aristotle[1]

E.1 Introduction: The Systematic Correction

Aristotle's philosophy represents the first systematic attempt to correct Plato's inversion of the General Zero Principle. Where Plato made the contrasting ground determinate (the Forms), Aristotle returned it to indeterminacy—not as Anaximander's boundless *Apeiron*, but as **potentiality** (*dynamis*, δύναμις). In doing so, he formalized the relational logic of contrast into a coherent metaphysical framework. Yet, by remaining committed to substance (*ousia*, οὐσία) as primary, he stopped short of the full Diachronic Turn that Neo-Pre-Platonic Naturalism completes. While substance-thinking was indeed characteristic of Plato's metaphysics, Aristotle radicalized and systematized it into the cornerstone of his own ontology. This appendix examines Aristotle as the thinker who gave GZP its first rigorous formulation, collected the empirical data that demanded a

---

1. Aristotle, *Physics* III.4, 204a2–6. NPN interprets Aristotle's struggle with the infinite here as the inevitable friction between a substance-based ontology and the indeterminate *Apeiron* required by GZP. In *The Complete Works of Aristotle: The Revised Oxford Translation*, ed. Jonathan Barnes (Princeton, NJ: Princeton University Press, 1984), 1:355.

process ontology, yet remained confined within the substance-paradigm that ultimately limited his vision.[2]

E.2 Formalizing the Ground: Potentiality and Actuality

The cornerstone of Aristotle's metaphysical system is his distinction between potentiality and actuality—a conceptual framework that directly formalizes the relational logic at the heart of the **General Zero Principle (GZP)**.

E.2.1 The Potentiality/Actuality Distinction

Aristotle's core metaphysical innovation is the distinction between potentiality and actuality (*energeia*, ἐνέργεια).

- **Potentiality:** The capacity to become, the field of unrealized possibilities.
- **Actuality:** The realized state, the determinate form or activity.

In NPN terms, this is the **GZP formalized**:

Potentiality (indeterminate field) → Actuality (determinate realization)

Aristotle applies this schema universally:

- **Physics:** Matter (potential) → Form (actual)
- **Biology:** Acorn (potential) → Oak tree (actual)
- **Psychology:** Capacity to reason (potential) → Exercised reason (actual)
- **Ethics:** Capacity for virtue (potential) → Virtuous activity (actual)

E.2.2 Correcting Plato's Inversion

Where Plato located the ground in a separate realm of perfect actualities (Forms), Aristotle correctly relocated it to **immanent potentiality**. The

---

2. For a comprehensive analysis of Aristotle's relationship to his predecessors, see W. K. C. Guthrie, *A History of Greek Philosophy, Vol. 6: Aristotle: An Encounter* (Cambridge: Cambridge University Press, 1981). Guthrie frames Aristotle as the great systematizer who naturalized the insights of the Academy.

APPENDIX E. ARISTOTLE AND THE ZERO PRINCIPLE 219

form of a statue is not in a transcendent Platonic heaven; it is the actualization of the marble's potential. This is a decisive return to the spirit of Anaximander: the ground is not a "more real" thing elsewhere, but the inherent capacity for determination here.[3]

E.3 The Biological Turn: Mapping the Results of Process

Aristotle's exhaustive biological research—his classification of species, analysis of reproduction, and study of animal motion—was not a mere empirical hobby. It was an **attempt to reverse-engineer cosmic process from its frozen results**. Lacking the concept of deep time or evolutionary change, he could only interpret the dynamic interplay of *Philia* (inheritance) and *Neikos* (variation/selection) as static "natures" or essences.[4]

E.3.1 Taxonomy as Fossil Record

When Aristotle grouped organisms by shared traits, he was mapping the historical work of *Philia*—the cohesive force that preserves structure across generations. When he noted diagnostic differences between species, he was recording the work of *Neikos*—the differentiating force that drives speciation. His *Scala Naturae* (Great Chain of Being) was an intuitive, if static, representation of emergent complexity.

E.3.2 The Limits of Substance-Thinking

Because Aristotle began with substance as primary, he could only interpret these diachronic patterns as synchronic essences. A species' "nature" (*physis*) was for him a fixed, timeless form rather than a dynamic, historical trajectory. He captured the **products** of the *Logos* but missed the **process** itself.[5]

---

3. For Aristotle's critique of the Forms as unnecessary duplications, see Aristotle, *Metaphysics*, 990a33–993a10. NPN views this as the necessary "Grounding" of the GZP: determination must be immanent.

4. See Aristotle, *History of Animals* and *Parts of Animals*. NPN interprets Aristotle's taxonomy as a "Synchronic Snapshot" of a Diachronic Process (Evolution).

5. For the seminal critique of substance ontology in favor of process, see Alfred North Whitehead, *Process and Reality* (New York: Free Press, 1978), 39–41. NPN adopts Whitehead's critique but rejects his "Eternal Objects," grounding process solely in the *Archē*.

E.4 The GZP Critique of the Unmoved Mover

From a GZP perspective, Aristotle's Unmoved Mover is a necessary but flawed solution. It correctly seeks a non-regressive foundation, but by making that foundation *pure actuality*, it creates a metaphysical singularity with no "outside." Pure actuality is, in its own way, just as determinate as Plato's Forms—it is the end point of all determination, with nothing beyond it.

GZP suggests a cleaner, more consistent solution: the ultimate ground cannot be characterized positively at all—not as "pure actuality," not as "pure potentiality," not as any determinate concept. To label it "pure potentiality" would be to violate FP5 (Impotence Before the *Apeiron*), which states that the *Apeiron* is the category for which empirical observation is impossible and about which no logical operation can be empirically grounded.

The *Apeiron* is **not a positive concept**—it is the **unknowable complement**, the necessary "outside" stipulated by the Zero Principle. Aristotle's Unmoved Mover, as pure actuality, remains a *something*—a final, determinate terminus. GZP, in contrast, requires that the first principle be **nothing**—the indeterminate ground against which the determinate *Archē* can exist. This is not a retreat into mystery, but a rigorous acknowledgment of the boundary where determination ends and the logic of contrast itself finds its limit.

E.4.1 Why Aristotle Could Not Take This Step

Aristotle's substance ontology prevented him from positing pure potentiality as the ultimate ground. For Aristotle, potentiality always belongs *to a substance* (marble has the potential to be a statue). A potentiality without a subject was inconceivable within his framework. Thus, his first principle had to be an actual substance—the Unmoved Mover.[6]

---

6. See Aristotle, *Metaphysics*, 1071b12–22. The Unmoved Mover must be actual to initiate motion, but NPN argues this "First Cause" violates the logic of contrast by positing a determination without an indeterminacy.

APPENDIX E. ARISTOTLE AND THE ZERO PRINCIPLE

E.5 The Four Causes and the Relational Web

Aristotle's doctrine of the four causes—material, formal, efficient, and final—can be read as an attempt to map the relational network that constitutes any determinate entity. Each cause answers a different "why?" question, but together they describe how a thing emerges from its context:

- **Material Cause:** The potentiality (*hylē*, ὕλη) from which it arises
- **Formal Cause:** The determinate structure it actualizes
- **Efficient Cause:** The preceding actuality that triggers the change
- **Final Cause:** The *telos* toward which it strives

This is a holistic, relational model of determination that respects the interconnectedness implied by GZP. Yet, because Aristotle anchors this network in primary substances, the system remains static at its foundation.[7]

E.6 Aristotle's Legacy: The Data and the Framework

Aristotle's philosophical achievement is twofold: he provided the first rigorous formalization of the General Zero Principle through his distinction between potentiality and actuality, and he assembled a vast empirical catalog that would later demand the very process ontology his substance-based system could not accommodate. His legacy to NPN is therefore not one of mere historical influence, but of indispensable tools and evidence that point toward the system's completion.

E.6.1 What Aristotle Gave NPN

Aristotle's contribution to the Neo-Pre-Platonic Naturalist framework is both foundational and preparatory. His work provides not only conceptual tools but also the empirical groundwork that points beyond his own system.

---

7. For a modern re-interpretation of the Four Causes as emergent dynamics, see Terrence W. Deacon, *Incomplete Nature: How Mind Emerged from Matter* (New York: W. W. Norton & Company, 2012), 29–45. Deacon's "Absential" causes align with the NPN view of the *Apeiron* as a necessary negative space.

1. **The Formal Framework:** The potentiality/actuality distinction as a precise formulation of GZP.
2. **The Empirical Data:** His biological works provide a monumental catalog of the results of cosmic process—a snapshot of deep-time dynamics frozen in taxonomic form.
3. **The Relational Model:** The four causes as a proto-systemic account of determination, mapping the web of relations that constitute any entity.
4. **The Explicit Rejection of Regress:** His rigorous arguments against infinite causal chains in *Metaphysics* XII prepare the logical ground for GZP as the non-regressive solution.[8]

E.6.2 Where Aristotle Fell Short

1. **Substance-Primary Ontology:** He never made the Diachronic Turn to process-primary thinking.
2. **Static Essences:** He interpreted historical patterns as timeless natures.
3. **The Unmoved Mover:** He posited a determinate first cause rather than an indeterminate ground.

E.7 Conclusion: Aristotle as Bridge and Limit

Aristotle stands as the great systematizer of the insight that Anaximander glimpsed and Plato inverted. He gave the General Zero Principle its first precise formulation (potentiality/actuality) and collected the empirical evidence that would later demand a process ontology. Yet, he remained within the substance-paradigm he inherited from Plato, unable to take the final step toward a fully diachronic metaphysics.

Neo-Pre-Platonic Naturalism does not reject Aristotle; it **completes him**. It takes his potentiality/actuality distinction and applies it to a process-primary reality. It takes his biological data and reinterprets it through

---

8. Aristotle, *Metaphysics* XII.7, 1072a20–25. Aristotle proves the necessity of a stop; NPN simply changes the nature of the stop from "Prime Mover" to "Indeterminate Ground."

APPENDIX E. ARISTOTLE AND THE ZERO PRINCIPLE

deep time and evolutionary dynamics. It accepts his demand for a non-regressive foundation but locates it in the *Apeiron* rather than the Unmoved Mover.

Aristotle was not wrong—he was **incomplete**. He built the most sophisticated substance-based system possible, one that contained within it both the formal expression of GZP and the empirical clues pointing beyond substance-thinking. In returning to the Pre-Platonic emphasis on Becoming while retaining Aristotle's analytical rigor, NPN achieves what Aristotle aimed for but could not reach: a metaphysics grounded in the indeterminate, explained through the dynamic, and verified by the historical record of the cosmos itself.

See also: Appendix A for the General Zero Principle
See also: Appendix D for Plato's application and inversion of GZP

Appendix F

The Confidence Gradient: From the Tyranny of Certainty to the Power of Probability

"No man knows, or ever will know, the clear truth about the gods and about all the things I speak of; for even if one should chance to say what is exactly the case, nevertheless he himself does not know it; but opinion is fashioned over all things."
— Xenophanes[1]

F.1   Introduction: The Allure and Peril of the Binary

The human mind, with its *Somatic logos*, has a deep-seated drive for clarity. This drive, an evolutionary adaptation for efficient decision-making, often manifests as a craving for **binary states**: true/false, known/unknown, certain/uncertain. This binary mode is the cognitive equivalent of a simple on/off switch—highly effective for immediate threats and opportunities, but catastrophically ill-suited for modeling a complex, dynamic, and layered reality.[2]

---

1. Xenophanes of Colophon, fragment B34, in *The Presocratic Philosophers: A Critical History with a Selection of Texts*, eds. G. S. Kirk, J. E. Raven, and M. Schofield, 2nd ed. (Cambridge: Cambridge University Press, 1983), 169–70.

2. For the cognitive science of heuristic thinking and the mind's reliance on simplifying binaries for efficiency, see Daniel Kahneman, *Thinking, Fast and Slow* (New York: Farrar, Straus and Giroux, 2011), 79–85.

The history of philosophy and ideology is littered with the wreckage of systems that succumbed to this craving, mistaking the map for the territory by enshrining a provisional model as certain dogma.[3] The Socratic *Elenchus* was designed precisely to shatter this false certainty, to induce the *Aporia* that is the necessary ground for growth. Yet, this dissolution of dogma is often misinterpreted as an invitation to nihilism—the belief that if no knowledge is certain, then no knowledge is reliable, and thus no action is meaningful.

This appendix argues that the NPN concept of a **Confidence Gradient** is the master key that unlocks a third path between the twin abysses of **dogmatic certainty** and **nihilistic despair**. It is not a compromise, but a superior epistemic framework that describes how we naturally, and successfully, function in the world.

F.2 The Functional Anatomy of the Confidence Gradient

The Confidence Gradient is not a mere admission of uncertainty; it is a **dynamic, multi-level assessment system** that the *Nous* uses to allocate cognitive resources and guide action. It can be visualized not as a simple sliding scale, but as a nested set of criteria:

- **Level 1: Predictive Power (The Engine of Utility):** This is the most fundamental test. A model gains confidence by consistently generating successful predictions when tested against new data from *Aisthēsis*. Its "truth" is its functional utility in navigating *Physis*. The law of gravity may not be a transcendent, certain truth, but its predictive power is so immense that acting on any other model would be immediately and catastrophically maladaptive. Our confidence in it is functionally absolute for all terrestrial navigation.

---

3. This systematic rejection of the demand for objective certainty aligns with Nietzsche's Perspectivism, the thesis that all knowledge is necessarily colored by the perspective of the striving agent rather than corresponding to a static "thing-in-itself." See Friedrich Nietzsche, *Beyond Good and Evil*, trans. Walter Kaufmann (New York: Vintage, 1966), §§ 34–36.

- **Level 2: Internal Coherence (The Architecture of the Model):** A model must be logically consistent with itself and with other high-confidence models. A hypothesis that requires suspending the laws of logic or fundamental physics to work carries a cripplingly low confidence from the start. The *Somatic logos* acts as a built-in integrity check.

- **Level 3: Explanatory Scope (The Breadth of the Map):** Confidence increases when a model can explain a wide range of seemingly disparate phenomena. Darwin's theory of evolution gained immense confidence not from a single fossil, but from its power to coherently explain biogeography, embryology, comparative anatomy, and the fossil record under a single, elegant framework.

- **Level 4: Fecundity (The Capacity for Growth):** High-confidence models are generative. They open new lines of inquiry, suggest novel experiments, and lead to further discoveries. A model that is a conceptual dead end, that explains its own narrow domain but nothing else, remains at a lower confidence tier.

This gradient is not a ladder to certain truth, but a **continual process of validation**. A model exists in a state of dynamic equilibrium, its confidence level constantly being adjusted by its ongoing interaction with the *Archē*. These criteria align with long-observed virtues of robust scientific theories, which are valued for their predictive power, internal consistency, breadth, and fertility.[4]

F.3 The Destruction of Certainty is the Acquisition of Power

The move from a binary (Certain/False) to a gradient (0% confidence → maximum confidence) is not a loss but a profound liberation. It replaces a brittle, static epistemology with a resilient, dynamic one.

---

[4]. For a classic analysis of the criteria for evaluating scientific theories, including explanatory power and coherence, see Carl G. Hempel, *Philosophy of Natural Science* (Englewood Cliffs, NJ: Prentice-Hall, 1966), 75–84.

- **The Tyranny of the Binary:** In a binary world, any crack in a model's certainty threatens total collapse. If a single piece of anomalous data can falsify a "certain" belief, the entire epistemic structure is fragile. This leads to **dogmatism**—the desperate, often violent, suppression of anomalies to protect the brittle certainty of the model. It also leads to the opposite pathology: when the model inevitably fails, the fall from "certain truth" to "certain falsehood" creates the vacuum that **nihilism** rushes to fill.

- **The Power of the Gradient:** In a gradient world, anomalous data is not a catastrophe; it is **information**. It causes a recalibration of confidence, not a total collapse. A scientist whose experiment contradicts a prevailing theory does not conclude that "all of science is a lie." They conclude, "My confidence in this specific aspect of the model must be reduced, and a new model with higher predictive power must be sought." This is the engine of scientific progress. It is a system designed for **learning**, not for **possession**.

The destruction of certainty is thus the destruction of a **cognitive idol**. It is the recognition that the feeling of absolute certainty is often a warning sign of a closed mind, not a mark of true understanding.

F.4 From Nihilism to Navigational Vigor

The charge of nihilism arises from a fundamental confusion: it mistakes the **absence of metaphysical certainty** for the **absence of functional reliability**.

- **Nihilism says:** "Nothing is certain, therefore nothing can be known, therefore nothing matters."
- **The Confidence Gradient says:** "Nothing is *metaphysically certain*, but everything (within the *Archē*) is *practically knowable* to a degree of confidence that makes action not only possible but mandatory. What matters is the continual process of alignment."

The Navigator on the open ocean has no *certain* knowledge of what lies over the horizon. The map is not the territory. The weather may change.

Yet, the navigator is not paralyzed. They act with high confidence based on their charts, their instruments, and their understanding of celestial mechanics and ocean currents. Their "knowledge" is a set of high-confidence models that are *reliable enough* to guide a successful journey.

This is not a "leap of faith" in the nihilist's sense. It is a **rational commitment based on accumulated evidence of a model's functional utility**. The *Hormē* demands action, and the Confidence Gradient tells us *how* to act, not *whether* to act.

- **We do not have certainty that the sun will rise tomorrow;** we have a model of planetary motion that has proven 100% predictively successful for every tomorrow of our lives and for all recorded history. Our confidence is so high that to bet against it would be insanity. This is not nihilism; it is the pinnacle of rational grounding.

- **We do not have certainty in our relationships;** we have a high-confidence model of another person's character, built through countless interactions. To demand Cartesian certainty of their future loyalty is to destroy the possibility of trust, which is itself a gradient-based, rational risk.

The rejection of certainty is thus a call to **vigilance, engagement, and courage**, not to apathy. It forces the Navigator to remain present, to pay attention to feedback from the *Archē*, and to be ready to revise their course. It replaces the passive comfort of dogma with the active, generative struggle of alignment.

F.5 The Popperian Protocol: Falsification as Navigation

The logic of the Confidence Gradient finds its definitive methodological expression in the work of Karl Popper. His core insight—that scientific knowledge advances not through verification but through **conjectures and refutations (falsification)**—is not merely a useful rule of thumb. It is the necessary operating procedure for a Navigator within an *Apeiron*-bounded *Archē*.[5]

Popper's system is the practical implementation of the NPN epistemological stance:

1. **A 'Conjecture' is a Model of the *Archē*.** It is a bold claim about the structure of the *Logos*, a proposed map of a specific territory.

2. **'Falsification' is a Controlled *Lysis*.** A rigorous test is an attempt to find a piece of the "not-yet-known *Archē*" that contradicts the model. A successful falsification is a moment of cognitive *Lysis*—the model shatters, creating productive *Aporia*.

3. **'Corroboration' is the Confidence Gradient.** A theory that survives fierce attempts at falsification has not been proven "true." Its **confidence level has been raised.** It has proven its functional utility and predictive power across a wider domain of the *Archē*.

This leads to the ultimate, ontological criterion for demarcation, derived directly from the **First Principle of Knowledge**:

> **A theory that is not falsifiable is not a theory about the *Archē*; it is a claim about the *Apeiron*.**

Such a claim attempts to make a substantive statement about that which, by definition, provides no empirical premises and lies beyond the boundary of the *Logos*. It cannot be tested, refined, or broken by interaction with reality. It is, epistemologically, a form of intellectual *Doxa*—a story told about the silent void.

Popper gave science its method. Neo-Pre-Platonic Naturalism provides the first-principles foundation for *why* that method is not just effective, but is the only one logically possible for a finite Navigator. The demand for falsifiability is the demand that our models be about the world we can navigate, and not the boundless context we cannot.

---

5. For Popper's core thesis on falsification, see Karl Popper, *The Logic of Scientific Discovery* (London: Routledge, 1982), esp. chap. 1.

APPENDIX F. THE CONFIDENCE GRADIENT

Theorem

T1: THEOREM: THE POPPERIAN PROTOCOL

Statement: The complete methodology of conjecture and refutation (falsification) is the necessary and optimal procedure for a finite *Navigator* to increase the Confidence Gradient of its models of the *Archē*. It is the systematic implementation of the epistemic cycle of *Lysis* and *Aporia*.

Derivation: This protocol is logically entailed by the First Principles of Neo-Pre-Platonic Naturalism.

1. The *Archē* exists as an objective reality (Primacy of the *Archē*), and a finite *Navigator* has emerged within it (Potential for *Nous*). The *Navigator*'s constitutive task is therefore to build functional models of this reality.

2. However, the *Archē* is fundamentally a process, not a static state (Diachronic Primacy). Consequently, no model can be a certain or final snapshot of reality.

3. It follows that knowledge must be a Confidence Gradient—a measure of a model's predictive success and utility—rather than a state of certain possession (Corollary: The Confidence Gradient).

4. Critically, the *Archē* is bounded by the *Apeiron*, the permanently unknowable (Impotence Before the *Apeiron*). This guarantees a permanent frontier of the "not-yet-known *Archē*" that can potentially contradict any current model.

5. Therefore, the only way to increase a model's Confidence Gradient is to test it aggressively against this frontier. A test that a model fails induces a cognitive *Lysis* and productive *Aporia*, clearing the ground for a better model. A test a model survives raises its justified confidence. This process of testing—of attempted refutation—*is* the logic of falsification.

Conclusion: Falsification is not merely a useful heuristic. It is the necessary and defining activity of a finite *Navigator* within a diachronic *Archē*. It is the engine of the Socratic *Elenchus* applied to the cosmos, and the only non-arbitrary procedure for the growth of knowledge.

F.6 A Scientific Exemplar: Diachronic Primacy and The Uncertainty Principle

The **First Principle of Diachronic Primacy**—that being is a stabilized pattern within Becoming—finds a clear exemplar in the operational *Logos* of the quantum realm. This view, which prioritizes process and relation over static substance, finds a profound resonance in the tradition of process philosophy.[6] The Heisenberg Uncertainty Principle ($\Delta x \Delta p \geq \hbar/2$) can be understood as a physical expression of this primacy.

The principle demonstrates that a perfectly synchronic snapshot of a particle's state—a definitive position and momentum at a single instant—is ontologically forbidden by the *Archē*.[7] The very attempt to force such a snapshot is a category error. The mathematical constraint ($\Delta x \Delta p \geq \hbar/2$) is the *Logos* enforcing its fundamental diachronic nature, revealing that what we call a "particle" is a process whose being cannot be reduced to a static point.[8]

F.7 Formal Derivation: The Necessity of the Confidence Gradient

This section presents a deductive proof showing that the Confidence Gradient follows necessarily from the First Principles of NPN. The necessary preconditions for this derivation include the fact that the emergent Nous is bounded by the very laws and limitations of the Archē that created it, establishing the status of the finite knower.

Premises (First Principles & Axioms)

---

6. For the foundational text of process philosophy, which argues that "Becoming" is ontologically primary and that reality is constituted by events and processes, see Alfred North Whitehead, *Process and Reality* (New York: The Free Press, 1978), 18–21, 23–26.

7. Heisenberg discusses the uncertainty relation not merely as a measurement limit, but as an epistemic gradient tied to the nature of reality. See Werner Heisenberg, *Physics and Philosophy: The Revolution in Modern Science* (New York: Harper & Row, 1958), 58–62.

8. The principle was formulated by Werner Heisenberg. See W. Heisenberg, "Über den anschaulichen Inhalt der quantentheoretischen Kinematik und Mechanik," *Zeitschrift für Physik* 43 (1927) 172–98. English translation: "The Physical Content of Quantum Kinematics and Mechanics," in *Quantum Theory and Measurement*, eds. J. A. Wheeler and W. H. Zurek (Princeton: Princeton University Press, 1983), 62–84.

APPENDIX F. THE CONFIDENCE GRADIENT

- **P1 (Primacy of the *Archē*):** An objective, lawful domain exists that allows empirical interaction.
- **P2 (Diachronic Primacy):** Being is a stabilized pattern within Becoming. Any fixed, synchronic description is therefore only an abstraction of a deeper, unfolding process.
- **P3 (The Finite Knower):** The *Nous* is a finite subsystem within the *Archē*, with bounded cognitive and observational capacity.
- **P4 (The *Apeiron* Boundary):** The *Archē* is bounded by the *Apeiron*, a category that cannot be observed or fully determined. This establishes a permanent epistemic horizon: there will always be aspects of reality that exceed possible observation or representation.

Definitions

- A **Model (C)** is any finite claim about the structure or behavior of the *Archē*.
- **Justification, J(C)**, is the measure of a model's validity or epistemic warrant. It is the criteria by which a Navigator accepts or rejects a model.

Derivation

1. **Models are inherently partial.** From the principle of diachronic primacy, every model is a simplified snapshot of an ongoing process. No model captures the full, unfolding reality it describes.

2. **Finite knowers face an open-ended reality.** Because the knower is finite, and because the *Apeiron* establishes an empirical boundary, the possible future states of the *Archē* will always extend beyond what any model can account for. There will always be potential observations that a given model does not anticipate or resolve.

3. **Absolute certainty is impossible for any model.** Since a model is always partial (Step 1), and since reality always contains possible future states the model cannot settle (Step 2), no model can ever be absolutely, finally, or metaphysically certain. Every model remains open to refinement, extension, or revision.

4. **Justification is necessarily graded.** Since absolute binary validation is impossible (from Step 3), the only remaining available metric for $J(C)$ is functional adequacy (defined as predictive success, coherence, and scope). Unlike binary truth, functional adequacy is an analog quality—a model can be slightly more predictive or slightly more coherent than another. Therefore, $J(C)$ *must* manifest as a gradient.

5. **A binary epistemology is incompatible with the First Principles.** Since no model can achieve final certainty, and since justification varies in strength, an epistemology that divides beliefs into either "certain" or "false" does not fit the structure of our situation as finite knowers. Instead, the only coherent epistemic framework is one that recognizes degrees of justification.

Conclusion

The Confidence Gradient is not an optional interpretive stance; it is a necessary consequence of our metaphysical situation.

> For any finite knower within a diachronic *Archē* bounded by the *Apeiron*, the only logically consistent epistemology is one grounded in a Confidence Gradient.

APPENDIX F. THE CONFIDENCE GRADIENT

> **Theorem**
>
> ### T2: THEOREM: THE STATUS OF FORMAL TRUTHS
>
> **Statement:** Analytic truths (logic, mathematics) are certain within their defined contrast-domains because their boundaries are stipulated. Synthetic knowledge—modeling the *Archē*—is gradient-bound because its boundary is the *Apeiron*, which cannot be stipulated or observed.
>
> **Derivation:**
> 1. By **ZP**, identity requires contrast.
> 2. In analytic systems, contrast is stipulated and controlled → **certainty**.
> 3. The *Archē* is contrasted by the *Apeiron* (**FP4**)—uncontrolled and un-model-able.
> 4. Therefore, synthetic knowledge is gradient-bound (**C2**).
> 5. Applying analytic systems to the *Archē* treats them as synthetic models → subject to **C2**.
>
> **Conclusion:** Analytic truth is grammar-certain. Synthetic truth is map-provisional. The Navigator uses the first to refine the second, always acknowledging the boundary between defined contrast and given contrast—the *Apeiron*.

F.8 Conclusion: Certainty is a Cage, The Gradient is the Compass

The Neo-Pre-Platonic Naturalist embrace of the Confidence Gradient is the final, mature step in the **Socratic** project. Socrates dismantled the false certainties of his contemporaries not to leave them in a void, but to prepare them for a more authentic relationship with the *Logos*.

To live by the Confidence Gradient is to accept our role as *Navigators* within the silent *Apeiron*. It is to understand that our models are tools, not treasures; maps, not territories. The goal is not to possess a final, static truth, but to participate in the endless, open-ended process of refining our maps, of converting the unknown into the known, and of steering our individual and collective vessels with ever-greater skill and grace through the dynamic, lawful, and wondrous *Archē*.

The destruction of certainty is not the death of meaning; it is the birth of the journey. And the Confidence Gradient is the compass that makes the journey possible.

Appendix G

The Navigator Protocol: Logical Mapping to First Principles

"To know that you do not know is the best. To think you know when you do not is a disease." — Laozi[1]

G.1 Introduction: The Logic of Practice

The First Principles describe what is. The Navigator Protocol describes what to do about it. This appendix demonstrates that every stage of human transformation—from cognitive crisis to flourishing—is not a collection of helpful suggestions but a strict logical derivation from the structure of reality itself.

To violate the Protocol is not 'sinful'; it is inefficient—a failure to align with the *Logos* that will manifest as friction, failure, and entropy.

The Protocol follows a precise 11-stage developmental arc, moving from the precondition of openness to the final state of flourishing.

G.2 Phase I: The Diagnosis (Breaking the Shell)

G.2.1 1. Eukairia (Receptive Moment)

Definition: The precondition for inquiry. A state where the *Thymos* is not defensive and the *Logistikon* is curious.

- Primary Mapping: FP9 (The Modeling Mandate)

---

1. Laozi, *Dao De Jing*, ch. 71.

- **Logical Argument:** The *Modeling Mandate* requires data uptake. If the *Thymos* (defense mechanism) is blocking input to protect *Doxa*, modeling cannot occur. *Eukairia* is the functional state of "Input Open / Defense Low" required to detect error.

G.2.2 2. Elenchus (Diagnostic Test)

Definition: A rigorous, systematic interrogation of belief structures to expose contradictions within the current model.

- **Primary Mapping: FP3 (Neikos) and GZP (General Zero Principle)**
- **Logical Argument:**
    1. **The Force (FP3):** *Neikos* is the force of differentiation and dissolution. *Elenchus* is "Cognitive Neikos"—applying stress to the model to find weak points.
    2. **The Logic (GZP):** A determinate model must be consistent. *Elenchus* seeks the infinite regress or circularity that signals a violation of the GZP within the current belief.

G.2.3 3. Lysis (Dissolution)

Definition: The structural failure. The cognitive breaking down of foundational assumptions triggered by contradiction.

- **Primary Mapping: GZP (General Zero Principle)**
- **Logical Argument:**
    1. **The Necessity (GZP):** Determination requires an indeterminate ground.
    2. **The Event:** When *Elenchus* exposes the contradiction, the determination (the specific belief) loses its footing. *Lysis* is the unavoidable return of the "Marked State" to the "Unmarked State" (Indeterminacy).

G.2.4 4. Exaiphnes (The Rupture)

Definition: The sudden "Aha!" moment of profound disorientation; the realization that certainty was an illusion.

- **Primary Mapping: FP9 (The Modeling Mandate)**
- **Logical Argument:**
    1. **The Recognition:** This is the moment of meta-cognition. The Navigator realizes, "I am not the map; I am the mapper."
    2. **The Shift:** It is the sudden shift from identifying with the broken model (*Doxa*) to identifying with the process of modeling the *Logos*.

G.3 Phase II: The Reconstruction (Aligning with Reality)

G.3.1 5. Katharsis (Purification)

Definition: The conscious choice to release attachment to the falsified belief, clearing the *Psyche* for inquiry.

- **Primary Mappings: FP1 (Primacy of the *Archē*) and FP5 (Impotence)**
- **Logical Argument:**
    1. **The Reality (FP1):** The *Archē* ignores our preferences.
    2. **The Choice:** Holding onto a falsified belief is functionally maladaptive. *Katharsis* is the "Somatic Release"—aligning the *Psyche* with the logic of *FP1*. We purge the false because it is dangerous.

G.3.2 6. Aporia (The Void)

Definition: A state of conscious ignorance where old paths are closed, forcing the mind to confront the unknown without a map.

- **Primary Mappings: C2 (The Epistemic Context) and FP5 (Impotence Before the *Apeiron*)**
- **Logical Argument:**
    1. **The Truth (C2):** We exist on a Confidence Gradient.
    2. **The Experience:** *Aporia* is the honest experience of the lower bound of that gradient. It is the refusal to fake certainty, honoring the limits of **FP5**.

G.3.3 7. Dikē (Re-anchoring)

Definition: The acceptance of constraints. Realizing the *Archē* operates by impersonal laws regardless of one's wishes.

- **Primary Mappings:** FP1 (Primacy of the *Archē*) and FP3 (Exhaustive Polarity)
- **Logical Argument:**
    1. **The Ground (FP1):** The new model must start from "What Is," not "What Should Be."
    2. **The Structure (FP3):** "What Is" is structured by *Philia* and *Neikos*.
    3. **The Anchor:** *Dikē* is the calibration step—anchoring the *Logistikon* to the actual causal laws of the territory.

G.4 Phase III: The Execution (Action and Mastery)

G.4.1 8. Prohairesis (Choice)

Definition: The commitment to a mode of engagement—either Structural Analysis (*Demiourgos*) or Synthesis (*Genetor*).

- **Primary Mappings:** FP2 (Diachronic Primacy), FP3, FP9
- **Logical Argument:**
    1. **The Flux (FP2):** Reality changes; therefore, the method must adapt.
    2. **The Calculation:** *Prohairesis* is the calculation of the correct vector. Do I need to build/maintain (*Demiourgos*) or disrupt/create (*Genetor*)?
    3. **The Act:** It is the application of **FP9** to select the future trajectory.

G.4.2 9. Energeia (Being-at-Work)

Definition: The active implementation. Sustained expression of the *Nous* deploying the new model as a living hypothesis.

APPENDIX G. THE NAVIGATOR PROTOCOL

- **Primary Mappings:** FP6 (Causal Continuity) → FP1
- **Logical Argument:**
    1. **The Chain (FP6):** Outcomes require causal input.
    2. **The Test:** *Energeia* is the deployment of the *Prohairesis* into the causal fabric of **FP1**. It transforms the hypothesis into physics.

G.4.3 10. Phronēsis (Practical Wisdom)

Definition: Automated alignment. The state where the *Logistikon* governs action in alignment with the *Logos* without requiring the active intervention of the *Nous*.

- **Primary Mappings:** All FPs (Integrated)
- **Logical Argument:**
    1. **The Habit:** Through repeated *Energeia*, the causal pathways of the *Somatic logos* are reinforced.
    2. **The Efficiency:** The system moves from "Conscious Competence" to "Unconscious Competence." The Navigator flows with the *Logos* automatically.

G.4.4 11. Eudaimonia (Flourishing)

Definition: The objective result. A life well-navigated, where the *Psyche* operates in harmony (*Dikaiosynē*) with reality.

- **The Equation:** $(FP9 + FP6) \rightleftharpoons (FP1 + FP5 + C2)$
- **Logical Argument:**
    - When the internal machinery is perfectly tuned to the external constraints, friction is minimized, and output is maximized. This objective functional state is *Eudaimonia*.

G.5 The Cybernetic Loop: The Popper Protocol

While the Navigator Protocol is presented linearly, its operational reality is cyclical. The mechanism that drives the Navigator from one stage to another—specifically the reaction to failure during *Energeia*—is the **Popper Protocol** (see Section F.5 for the full derivation).

This protocol defines the specific function of the *Nous* when a model encounters friction with the *Archē*.

G.5.1 The Mechanism of Correction

When the Navigator is in **Energeia** (Step 9) and the deployed model produces a prediction error (a mismatch between intent and result), the *Nous* is triggered out of automaticity (*Phronesis*) and back into conscious processing.

The *Nous* then executes one of two vector corrections based on the severity of the error:

1. The Minor Loop (Realignment)

Condition: The error is a calibration issue; the core structure of the model remains sound, but the parameters are off. * **Action:** The *Nous* applies **Cognitive Neikos** to trim the error. * **Trajectory:** The Navigator adjusts the *Prohairesis* (Step 8) and re-engages *Energeia* (Step 9). * **Result:** Refinement of the existing map.

2. The Major Loop (The Return to Aporia)

Condition: The error is structural; the map effectively falsified the territory. The prediction failure is catastrophic to the model's logic. * **Action:** The *Nous* accepts the falsification (**FP9**). * **Trajectory:** The Navigator abandons *Energeia* and triggers **Lysis** (Step 3), returning to the **Aporia** (Step 6) to build a fundamentally new model that incorporates the contradictory information. * **Result:** Evolution of the Navigator.

> **Note:** This feedback loop is the operational definition of **C2 (The Confidence Gradient)**. The Navigator does not seek "final truth"; they seek the continuous reduction of error through the Popperian cycle of Conjecture and Refutation.

G.6 Summary Table: The 11 Stages of the Navigator

Table G.1 The Navigator Protocol Mapped to First Principles

Step	Concept	Functional Definition & Mapping
1. *Eukairia*	Receptive Moment	Precondition for inquiry (FP9); *Thymos* down, *Logistikon* up.
2. *Elenchus*	Diagnostic Test	Interrogation of belief via Cognitive Neikos (FP3) to find GZP violations.
3. *Lysis*	Dissolution	Structural failure of the old model; return to Indeterminacy (GZP).
4. *Exaiphnes*	The Rupture	Sudden realization of the self as Modeler (FP9); the meta-cognitive break.
5. *Katharsis*	Purification	Releasing attachment to Doxa; prioritizing Reality (FP1) over comfort.
6. *Aporia*	The Void	Conscious ignorance (FP5); confronting the unknown without a map.
7. *Dikē*	Re-anchoring	Acceptance of constraints (FP1/FP3); grounding the mind in impersonal law.
8. *Prohairesis*	Choice	Commitment to engagement (Demiourgos/Genetor) based on Diachronic flux (FP2).
9. *Energeia*	Being-at-Work	Active implementation; deploying the model into the Causal Chain (FP6).
10. *Phronēsis*	Practical Wisdom	Automated alignment; the *Logistikon* governs without active *Nous* intervention.
11. *Eudaimonia*	Flourishing	The objective result; Systemic Harmony of the Psyche with Reality (FP1).

G.7 Conclusion: The Architecture of Alignment

The Navigator Protocol is not a collection of arbitrary moral commandments, nor is it a self-help heuristic for subjective well-being. It is the **logically necessary operational procedure** for a finite, model-building system (the *Nous*) attempting to persist within a lawful, dynamic reality (the *Archē*).

If the First Principles describe the "physics" of the cosmos, this Protocol describes the "engineering" of the competent agent. Its necessity is derived directly from the structural requirements of existence:

G.7.1 The Ethical Imperative of Accuracy (Theorem T4)

Central to this necessity is **Theorem T4: Ethical Isomorphism**. In the NPN framework, the distinction between "finding the truth" (epistemology) and "doing the good" (ethics) is revealed as a synchronic illusion.

- **The Identity:** Since "Good" is defined as action that successfully fulfills the *Hormē* within the constraints of the *Logos*, and "Truth" is the alignment of the internal model with that same *Logos*, they are functionally identical.

G.7.2 The Resolution of the Paradox: Truth-Survival Isomorphism

Any system that denies the isomorphism between Truth and Survival collapses into an immediate, lethal contradiction. If "Truth" were distinct from "Survival," we would be forced to accept two absurdities:

1. **The Benefit of Delusion:** It would be functionally "Good" for an agent to be "False" (deluded), as error would provide a competitive advantage over accuracy.
2. **The Lethality of Truth:** Truth would become a functional detriment—a weapon that reality uses to dissolve any agent "honest" enough to perceive it.

If a Navigator possesses a model that identifies a cliff as a plain, the "Truth" of gravity is not an abstract concept; it is an entropic force that destroys the agent for its lack of alignment. Therefore, the *Hormē* (the drive to persist) is fundamentally grounded in the truth of the *Archē*.

The "Ought" is revealed as nothing more than a localized expression of the "Is." To say an agent *ought* to act in a certain way is to say that such an action is the only one *isomorphic* with the truth of the environment. Virtue is not a moral additive; it is the state of being "Rightly Cut" (*Dikaiosynē*)—aligned so precisely with the grain of reality that the "penalty" of *Dikē* is avoided. Therefore, a system where Truth is self-detrimental is a system optimized for its own extinction. NPN asserts the only stable conclusion: **To flourish is to align with Reality.**

G.7.3 The Minimization of Friction

The Protocol functions as a system for minimizing the entropic friction that threatens any complex system:

- **Internal Harmony:** By subjecting *Doxa* to *Elenchus*, the Navigator resolves the civil war between the conscious mind's false models and the somatic wisdom of the body. This aligns the *Logistikon* with the *Hormē*, eliminating the internal drag of cognitive dissonance.
- **External Efficacy:** By grounding models in *Dikē* and testing them through *Energeia* (action), the Navigator minimizes the collision between their map and the territory. Action ceases to be a struggle *against* the *Archē* and becomes a navigation *of* it.

G.7.4 The Alignment of *logos* to *Logos*

Ultimately, the Protocol is the mechanism for synchronization. It is the process by which the internal *logos* (the subjective map) is continuously recalibrated to match the external *Logos* (the objective structure).

When this alignment is achieved, the gap between "Is" and "Ought" closes. The Navigator does not struggle to do what is right; through the cultivation of *Phronēsis*, the right action becomes the natural, path-of-least-resistance output of a healthy system. The Protocol converts the **Potential for Nous** (FP4) into the **Actuality of Eudaimonia.** It is the discipline of freedom—not freedom *from* reality, but the freedom to move effortlessly *within* it.

Appendix H

The Universal Grammar: The Daoist Mirror of the Presocratic System

"The great man is he who is in harmony with Heaven and Earth, with Yin and Yang, with the seasons and the spirits."
— I Ching (Book of Changes)[1]

"These two forces, Love and Strife, existed before, and shall be, nor ever shall boundless time be emptied of them."
— Empedocles[2]

H.1 Introduction:
A Shared Vocabulary for Reality's Rhythm

The recovery of the Pre-Platonic framework in the West finds a powerful and elegant parallel in the ancient Chinese conception of Yin and Yang. Where Heraclitus and Empedocles described the dynamic interplay of opposing forces, Chinese philosophy systematized this insight into a comprehensive cosmology. Neo-Pre-Platonic Naturalism posits that this is not a

---

1. *Great Treatise (Dazhuan)*, in *The I Ching or Book of Changes*, trans. Richard Wilhelm (Princeton: Princeton University Press, 1950), 292. NPN identifies this "harmony" as *Eudaimonia*—the functional alignment of the Navigator with the *Logos*.

2. Fragment 16, in *Die Fragmente der Vorsokratiker*, ed. Hermann Diels and Walther Kranz (Berlin: Weidmann, 1951), vol. 1, 318. NPN cites this to establish the **Universality of the Dual Engine**: *Philia* and *Neikos* are not cultural metaphors but objective cosmic constants.

coincidence, but evidence of different cultures mapping the same fundamental territory of the *Archē*, a view supported by rigorous comparative studies of early Greek and Chinese thought.[3]

This appendix argues that the alignment between these traditions extends far beyond a superficial resemblance of "opposing forces." It reveals a precise, tripartite structural isomorphism that validates the core axioms of NPN:

1. **The Pre-Ontological Ground:** Both systems recognize that determinate existence requires an indeterminate background (**Wuji / Apeiron**) to provide the contrast necessary for identity.
2. **The Constitutive Structure:** Both define reality not as static substance, but as a unified field of potentiality and principle (**Taiji / *Archē***).
3. **The Dynamic Engines:** Both identify the same two vectors of operation—Union and Separation (**Yin-Yang / *Philia-Neikos***)—as the drivers of all becoming.

We must see that both systems converge on the very same conclusions about reality: not because they shared a language, but because they faced the same *Logos*. To navigate the *Archē* is to grapple with these universal mechanics. To understand this alignment, we must begin where existence itself begins: with the necessary silence of the Zero Principle.

H.1.1 The Generative Sequence: Wuji, Adikia, and the Taiji

The structural isomorphism extends to the very origin of form. The Daoist progression from *Wuji* (Non-Polarity) to *Taiji* (Supreme Polarity) mirrors Anaximander's movement from *Apeiron* to *Adikia*.

- **Wuji is the Apeiron**: The indeterminate ground before distinction.

---

3. For a seminal comparative study of the conceptual frameworks of early China and Greece, see G. E. R. Lloyd and Nathan Sivin, *The Way and the Word* (New Haven: Yale University Press, 2002), 1–15. NPN interprets this convergence not as coincidence, but as independent verification of the *Archē's* structure.

APPENDIX H. THE UNIVERSAL GRAMMAR: DAOISM AND NPN

- **The Emergence of Taiji is Adikia**: The "Great Ultimate" is the first distinction—the separation of Yin and Yang. This parallels the "injustice" of a determinate form emerging from the boundless.
- **The Operation of Yin and Yang is Philia and Neikos**: Just as *Yin* (the receptive, cohesive principle) binds energy into form (*Philia*), *Yang* (the active, dispersive principle) drives the transformation and eventual dissolution of those forms (*Neikos*).

Both systems describe reality not as a static collection of things, but as a temporary, rhythmic disturbance in the void—a "hoarding" of form (*Yin/Philia*) that must eventually be paid back to the whole through change (*Yang/Neikos*).

H.1.2 *Wuji* (無極): The Ontological Silence of the Zero Principle

"The Dao that can be told is not the eternal Dao. The name that can be named is not the eternal name. The unnamable is the eternally real. Naming is the origin of all particular things." — Laozi, *Dao De Jing*, Chapter 1

The Daoist concept of **Wuji** (無極)—literally the "Ultimateless" or "limitless void"—represents the primordial state of non-distinction that precedes the emergence of the *Taiji* (the Supreme Ultimate).[4] In the metastructure of Neo-Pre-Platonic Naturalism (NPN), **Wuji** functions as the direct ontological equivalent of the **Zero Principle (ZP)**. It is the stipulative acknowledgment of the **Apeiron**: the necessary, indeterminate background that must be "marked" before the **Archē** (the figure) can be navigated.

The Differential Field

The Zero Principle dictates that identity is not intrinsic but differential. Just as a point on a graph requires the "0,0" coordinate as a reference

---

4. For the foundational definition of *Wuji* generating the *Taiji*, see Zhou Dunyi, *Taijitu shuo* (Explanation of the Diagram of the Supreme Polarity), in *Sources of Chinese Tradition*, vol. 1 (New York: Columbia University Press, 1999), 673–74. This sequence is the precise structural equivalent of the NPN **Zero Principle**: Determinacy (*Taiji*) requires an Indeterminate Ground (*Wuji*).

for its location, the primary polarity of *Yin* and *Yang* (correlating to **FP3: Exhaustive Polarity**) requires **Wuji** as the field in which to manifest.[5] Without this "Zero-point" of potentiality, the *Archē* collapses into a logical impossibility; it would be a "something" defined against nothing, which violates the relational requirements of the *Logos*. For the complete derivation of GZP and its Corollary ZP, see Appendix A.

Navigational Significance

For the **Navigator**, recognizing **Wuji** is not a mystical retreat but a technical necessity. It serves as the **Apeironic Context (C1)**: the reminder that all models (*Epistēmē*), however complex, are "cuts" made into a boundless indeterminate whole.[6] By acknowledging **Wuji**, the *Nous* remains agile, understanding that when a specific model fails (*Aporia*), the system does not end; it simply returns to the "Ultimateless" potential to be re-marked and re-navigated.

H.1.3 The Constitutive Layer: *Hylē* and *Logos* as Yin and Yang

The mapping of *Philia* and *Neikos* describes the **dynamics** of the system (how it moves). However, the NPN framework reveals a second, deeper layer of correspondence regarding the **constitution** of the *Archē* itself (what it is).

- **Yin as *Hylē* (The Receptive):** Just as *Hylē* is the dynamic potentiality or "hardware" of reality that receives structure, Yin is the dark, receptive, material principle (Earth). It is the substrate of existence that allows for manifestation.
- **Yang as *Logos* (The Creative):** Just as the *Logos* is the active, structuring principle or "software" that actualizes potential, Yang is the light, active, spiritual principle (Heaven). It is the informing

---

5. Zhou Dunyi explicitly frames this as a generative sequence: "The Supreme Polarity in activity generates *yang*... In stillness it generates *yin*... The Supreme Polarity is fundamentally Non-polar (*Wuji*)." See Zhou, *Taijitu shuo*, 673.

6. This aligns with the Zhuangzi's use of *Wuji* to denote the "boundless" reality that transcends human naming. See *Zhuangzi*, chap. 6. This validates **FP5**: the *Apeiron* cannot be named because it is the condition for naming.

APPENDIX H. THE UNIVERSAL GRAMMAR: DAOISM AND NPN

agency that generates determinate things out of the indeterminate substrate.[7]

The *Archē* as *Taiji*: In this view, the NPN *Archē*—the total, dynamic system of reality—finds its precise correlate in the ***Taiji*** (The Supreme Ultimate). The *Taiji* is the unified source where Yin (*Hylē*) and Yang (*Logos*) are not separate substances but co-constitutive aspects of a single reality.

Figure H.1: The Taiji (Supreme Ultimate): The Unified Dynamic of Logos (Yang) and Hylē (Yin)

H.1.4 The Dynamic Layer: The Engines of Interaction

If the *Taiji* describes the structure of the *Archē*, the **Two Forces** describe the engine that drives its evolution. Here, the cross-cultural mapping reveals the mechanism of the **Universal Dialectic** (FP3).

1. The Force of Union and Cohesion

- **NPN Term: *Philia* (Attraction).** *Philia* is the ontological vector of **binding**. It is the force that pulls disparate elements into unified

---

[7]. This structural parallel resonates with the Neo-Confucian metaphysics of Zhu Xi, who defined reality as the interplay of *Li* (Principle/Logos) and *Qi* (Matter-Energy/Hylē). See Zhu Xi, *Zhuzi yulei*, trans. Daniel K. Gardner, *Learning to Be a Sage* (Berkeley: University of California Press, 1990), 88–95.

wholes, creating structure, continuity, and "spheres" of stability.[8] In the NPN evolutionary framework, *Philia* manifests as **Inheritance**: the capacity of a system to replicate structure and maintain its identity across time against the dissolving flow of entropy. It is the conservative principle that preserves "what works."

- **Chinese Term:** *Yin* **(The Receptive).** *Yin* corresponds to **Earth** and the attribute of **Devotion**. It is not merely "passive" in a negative sense, but *Receptive*: it provides the necessary substance (*Hylē*) and spatial context in which the Creative impulse acts.[9] *Yin* is the sustainer of form, the "mother" of gravity that holds the ten thousand things in their distinct shapes.

- **Correlation:** Both *Philia* and *Yin* represent the **centripetal** tendency of reality—the drive to gather, hold, cohere, and sustain. They provide the **ontological mass** required for existence to persist.

2. The Force of Division and Action

- **NPN Term:** *Neikos* **(Repulsion).** *Neikos* is the ontological vector of **separation**. It is the force that breaks the static unity of the Sphere to create distinction, plurality, and movement.[10] In the NPN evolutionary framework, *Neikos* manifests as **Variation** and **Selection**: the drive to diverge from the norm, test new configurations, and cull what is no longer functional. It is the radical principle that drives innovation.

---

8. In Empedocles' cosmic cycle, *Philia* is the force that unites the four roots into the "Sphere" (*Sphairos*). See Kirk, Raven, and Schofield, *The Presocratic Philosophers*, 295–301. This correlates with the centripetal nature of *Yin*.

9. Wilhelm defines the Receptive (*K'un*) as the "perfect complement of the Creative," representing nature, space, and earth. NPN strips *Yin* of passive connotations; it is the **Structural Constraint** (*Hylē*) required for form. See Wilhelm, *The I Ching*, 10–11, 386–88.

10. For Empedocles, *Neikos* is the prerequisite for the existence of the cosmos; without separation, there would be no distinct entities. See Kirk, Raven, and Schofield, *The Presocratic Philosophers*, 302–5. This correlates with the centrifugal nature of *Yang*.

- **Chinese Term: *Yang* (The Creative).** *Yang* corresponds to **Heaven** and the attribute of **Movement**. It is the initiating power, the "Great Beginning" that generates time through action.[11] *Yang* is the fecundating spark—the aggressive vector that breaks inertia and forces potentiality into actuality.

- **Correlation:** Both *Neikos* and *Yang* represent the **centrifugal** tendency of reality—the drive to expand, act, differentiate, and individuate. They provide the **ontological energy** required for existence to evolve.

H.2 The Divergence: Structural Analysis vs. Process Navigation

While the metaphysical identification of the forces is nearly identical, the NPN analysis reveals a crucial divergence in methodological approach.

The Greek Turn: Mastery through Dissection (Cognitive *Neikos*)
The Greek tradition, particularly after Parmenides and Aristotle, harnessed the power of *Neikos* (Separation) as a cognitive tool. To understand the *Archē*, they broke it down. They developed atomism, logic, and categorical ontologies—mental blades designed to slice the seamless whole of nature into discrete, manageable parts.[12]

- **The Strength:** This created the foundation for Western science and the *Demiourgos* power of precise manipulation.
- **The Cost:** It led to the "Synchronic Flattening"—the tendency to mistake the static map for the dynamic territory.

The Chinese Turn: Mastery through Resonance (Cognitive *Philia*)
The Chinese tradition, conversely, maintained a focus on the *Logos* as a

---

11. Wilhelm describes the Creative (*Ch'ien*) as the power of "Time" that flows through the "Space" of the Receptive. NPN identifies *Yang* as the **Vector of Actuality** (*Logos*) that drives the process of Becoming. See Wilhelm, *The I Ching*, 4–6, 369–72.

12. For the Greek predilection for distinctness and logical separation, see G. E. R. Lloyd, *Polarity and Analogy* (Cambridge: Cambridge University Press, 1966). The Greek focus on *Neikos* (dissection) enabled science but risked the **Synchronic Flattening**.

flowing process (The Tao). Rather than dissecting the world to control it, they sought to resonate with it.

- **Confucianism** focuses on **Social Resonance**: *Li* (Ritual) is not a rigid law, but a dynamic choreography designed to keep the human community flowing in sync with the cosmic pattern.[13]
- **Taoism** focuses on **Natural Resonance**: *Wu Wei* (Non-Action) is not passivity, but the ultimate navigational efficiency—using the existing momentum of the *Archē* to achieve one's ends without wasting energy on friction (*Neikos*). "The sage acts without doing anything and teaches without saying anything."[14]

Table H.1   The Methodological Divergence: Structure vs. Flow

Domain	The Greek Turn (Structural)	The Chinese Turn (Processual)
Cognitive Mode	**Dissection (*Neikos*):** Breaking the whole into discrete, manageable parts (Atomism).	**Resonance (*Philia*):** Tuning the self to align with the flowing whole (Holism).
Primary Metaphor	**The Edifice:** Reality is a structure to be built, mapped, and maintained.	**The Stream:** Reality is a current to be navigated and flowed with.
Political Technology	**Rule of Law:** External architecture (constitutions) to contain and channel human nature.	**Rule of Virtue:** Internal cultivation (*Te*) to harmonize the ruler with the mandate.
Navigational Risk	**Synchronic Flattening:** Mistaking the static map for the dynamic territory.	**Vague Vitalism:** Lacking the precise models required to falsify errors.

---

13. Confucius, *Analects*, 2.3. *Li* is the technology of **Social Philia**—maintaining the cohesive web of society.

14. Laozi, *Tao Te Ching*, ch. 2. NPN redefines *Wu Wei* not as mysticism, but as **Navigational Efficiency**: maximizing output by aligning with the *Archē's* inherent momentum.

H.3 The Political Corollary: Architecture vs. Cultivation

This divergence explains the distinct political "technologies" of the two civilizations.

- **The West (Structural):** Developed the **Rule of Law**. This is a "Demiurgic" solution—building an external, static architecture (constitutions, checks and balances) to contain and channel human nature. It assumes the *Logos* must be imposed from the outside.
- **China (Processual):** Developed the **Rule of Virtue**. This is a "Navigational" solution. The "Mandate of Heaven" (*Tianming*) posits that order comes from the internal alignment of the ruler with the cosmic Tao.[15] If the Navigator-King is in *Energeia* (alignment), the society naturally harmonizes. If he loses that alignment, the flow turns chaotic (*Neikos*), and he loses the Mandate.

H.4 The NPN Synthesis: Merging East and West

NPN argues that the Western tradition essentially "forgot" the Tao (the dynamic process) in favor of the Atom (the static part). The "Great Divergence" discussed in Chapter 9 was a divergence away from this kind of Process Naturalism.

The complete Navigator requires the synthesis of these modes:

1. **Greek Precision:** We need the *Neikos* of the Western intellect to build high-fidelity models, falsify errors, and construct the "laws" of the *Second-Best State*.
2. **Chinese Fluidity:** We need the *Philia* of the Eastern wisdom to remember that these models are just maps of a flowing territory. We must navigate with *Wu Wei*—economical, timely action—recognizing that the ultimate reality is not a static building, but a moving stream.

---

15. *The Book of History* (*Shujing*), "The Announcement of Shao," and Mencius, *Mencius*, 1B.8. The "Mandate" is the feedback loop of **Dikē**: a ruler in misalignment with the *Logos* inevitably triggers the *Neikos* of rebellion.

H.5 Conclusion: The Universal Grammar of Identity

The striking convergence between Presocratic Naturalism and Daoist Metaphysics is not a mere historical curiosity; it is a validation of the **Diachronic Turn**. That two distinct civilizations, separated by vast geography and culture, arrived at an identical structural map of reality suggests that the *Logos* is not a cultural invention, but an objective discovery.

The Pre-Ontological Anchor Most crucially, this comparative analysis validates the **Zero Principle** as the fundamental law of existence. Both the NPN *Apeiron* and the Daoist *Wuji* assert the same radical truth: **Identity is differential, not intrinsic.**

- A "thing" cannot be a thing unless it stands out against a background of "no-thing."
- The *Archē* (The Bound) cannot exist without the *Apeiron* (The Boundless).
- The *Taiji* (The Pole) cannot manifest without *Wuji* (The Ultimate-less).

This universal recognition confirms that the definition of identity is **pre-ontological**: before any specific "stuff" (matter, energy, atoms) can exist, the logical necessity of a contrasting background must be in place. This effectively destroys the mystical notion of "self-existing essences" and grounds reality in a hard, logical structure of relation.

The Engine of Process Once this identity is established against the void, the operational grammar of the cosmos takes over. Whether we call the driving forces *Philia* and *Neikos* or *Yin* and *Yang*, the mechanism remains the same: reality is the eternal, oscillating tension between the drive to cohere (Union) and the drive to differentiate (Separation).

For the Neo-Pre-Platonic Naturalist, this cross-cultural isomorphism provides a profound confidence. Our model of the *Archē*—as ***Hylē*-as-actualized-by-*Logos***, bounded by the silent *Apeiron* and driven by *Exhaustive Polarity*—is not an arbitrary theoretical construct. It is a description of the "deep code" of the cosmos.

Appendix I

The Somatic Dhamma: NPN, Buddhism, and the Quieting of Internal Neikos

> *"Do not go upon what has been acquired by repeated hearing; nor upon tradition... nor upon a bias towards a notion that has been pondered over... When you yourselves know: 'These things are bad... these things lead to harm and ill,' abandon them. When you yourselves know: 'These things are good... these things lead to benefit and happiness,' enter on and abide in them."* – The Buddha[1]

I.1 Introduction: An Unexpected Convergence

On the surface, the traditions could not appear more different: the fiercely rational, world-engaging inquiry of the Pre-Platonics, and the introspective, renunciative path of the Buddha. Yet, when viewed through the functional, diachronic lens of Neo-Pre-Platonic Naturalism, a profound convergence emerges. This project of interpreting Buddhism through a naturalistic and scientific lens finds resonance with modern philosophical work.[2] Both are sophisticated systems aimed at diagnosing a fundamental human ailment and prescribing a functional cure. For NPN, the ailment

---

[1]. The Buddha, *Kalama Sutta* (AN 3.65), in *The Numerical Discourses of the Buddha: A Translation of the Anguttara Nikaya*, trans. Bhikkhu Bodhi (Boston: Wisdom Publications, 2012).

[2]. For a foundational work in the project of "naturalizing" Buddhism, interpreting its claims without supernatural metaphysics, see Owen Flanagan, *The Bodhisattva's Brain: Buddhism Naturalized* (Cambridge, MA: MIT Press, 2011), 1–6, 113–42.

is misalignment with the *Logos*; for Buddhism, it is *Dukkha*—often translated as suffering, stress, or unsatisfactoriness.[3]

This appendix argues that the core of the Buddhist path can be understood as a therapeutic technology for resolving a critical internal conflict: the war between the conscious mind's flawed models (*Doxa*) and the body's deep, implicit genetic knowledge—the recorded successes of *Philia* and *Neikos* that allowed one's specific ancestry to survive and replicate. The result of this war is **Internal Misalignment**, and the Buddhist path is a practice of surrender to a deeper truth, leading to *Dikaiosynē* (functional harmony) and consequent reduction of external friction with the *Logos* (lawful reality).

I.2 The Diagnosis:
Dukkha as a Rebellion Against One's Own Lineage

The Buddha's First Noble Truth is the declaration of *Dukkha*. From the NPN vantage, *Dukkha* is the somatic and emotional feedback of a *Psyche* in a state of rebellion against its own constitutive history.

- **Genetic Knowledge as *Somatic logos*:** Every human is born with a somatic inheritance—a deep-time record of what constituted successful navigation for their ancestors. This is not conscious belief, but a hardwired set of biases, drives, and alarm systems shaped by the perpetual dance of *Philia* (the forces that led to successful bonding, cooperation, and rearing of young) and *Neikos* (the forces that led to successful competition, boundary-setting, and avoidance of threats).[4] The "absences of potential ancestors" are the silent testimony to strategies that failed.

---

3. For the foundational teaching of the Four Noble Truths, of which *Dukkha* is the first, see *The Connected Discourses of the Buddha: A Translation of the Saṃyutta Nikāya*, trans. Bhikkhu Bodhi (Boston: Wisdom Publications, 2005), SN 56.11.

4. This concept is supported by the modern understanding of a complex, evolved human nature. See Steven Pinker, *The Blank Slate: The Modern Denial of Human Nature* (New York: Viking, 2002), 33–45.

APPENDIX I. THE SOMATIC DHAMMA: BUDDHISM AND NPN

- **The Conscious *Nous* as a Rebellious Agent:** The conscious, model-building *Nous* often operates on flawed *Doxa*—cultural conditioning, personal trauma, and ignorant desires. This conflict mirrors the modern cognitive model of the mind as containing automatic, intuitive processes that conscious reasoning often struggles to control.[5] It pursues goals (endless acquisition, ego inflation, avoidance of all discomfort) that are directly at odds with the parameters of success encoded in its own genetic substrate. To crave only pleasure and reject all pain is to declare war on the very alarm system that one's ancestors relied on for survival.
- ***Dukkha* is the Symptom:** This conflict—the conscious mind fighting the deep, somatic wisdom of the body—generates immense **Internal Misalignment**. The stress, anxiety, and pervasive unsatisfactoriness of *Dukkha* are the signals of a system trying to operate contrary to its own foundational, time-tested programming.

I.3 The Prescription: The Eightfold Path as a Protocol for Somatic Realignment

The Buddha's Noble Eightfold Path is not a set of metaphysical beliefs, but a functional guide for ceasing this rebellion. It is a method for dissolving the ignorant *Doxa* of the conscious mind so that the *Psyche* can fall back into alignment with its own *Somatic logos*.[6]

The path systematically retrains the *Psyche*:[7]

---

5. For the metaphor of the conscious mind as a rider trying to control a powerful, intuitive elephant, see Jonathan Haidt, *The Righteous Mind: Why Good People Are Divided by Politics and Religion* (New York: Pantheon Books, 2012), 3–6, 45–52.

6. For the detailed analysis and definition of the eight factors, see the *Vibhaṅga Sutta* in *The Connected Discourses of the Buddha*, SN 45.8.

7. For the classification of the eight factors into the three aggregates of Wisdom (*Paññā*), Virtue (*Sīla*), and Concentration (*Samādhi*), see the *Culavedalla Sutta* in *The Middle Length Discourses of the Buddha: A Translation of the Majjhima Nikāya*, trans. Bhikkhu Nanamoli and Bhikkhu Bodhi (Boston: Wisdom Publications, 1995), MN 44.

I.3.1 I. Wisdom (*Paññā*): Adopting the Correct Map

- **Right View:** This is the foundational *Epistēmē*: seeing reality as impermanent (*anicca*) and without a permanent, solid self (*anattā*). In NPN terms, this is the model that acknowledges the body as a transient process shaped by deep-time forces, not a vehicle for a separate, commanding ego. It is the intellectual acceptance that one's conscious desires are not the ultimate authority.
- **Right Intention:** This is the *Prohairesis* for harmony. It is the deliberate commitment to renounce the cravings and aversions that fuel the internal rebellion, and to cultivate actions that are non-harmful and compassionate—actions that resonate with the successful *Philia* strategies of one's lineage.

I.3.2 II. Ethical Conduct (*Sīla*): Ceasing Harmful Actions

- **Right Speech, Right Action, Right Livelihood:** These precepts are the *Demiourgos* framework for behavior. By consciously refraining from lying, stealing, and harming, one stops generating the external consequences and internal guilt that inflame the *Thymos* and agitate the *Orexis*. This creates a "ceasefire" in the external world, which is a prerequisite for quieting the internal war.

I.3.3 III. Mental Discipline (*Samādhi*): Disarming the Rebel Commander

- **Right Effort:** The conscious application of energy to cultivate skillful states and abandon unskillful ones. This is the execution of *Prohairesis* and *Energeia*.
- **Right Mindfulness (*Sati*):** The practice of observing the contents of consciousness—the cravings, the fears, the stories of the self—without identifying with them or acting on them. It is the practice of seeing *Doxa* as *Doxa*.
- **Right Concentration:** Unifies the mind, calming the chaotic strife of *Orexis* and *Thymos*, and allowing the *Logistikon* to stand down from its futile war.

APPENDIX I. THE SOMATIC DHAMMA: BUDDHISM AND NPN

Table I.1 The Noble Eightfold Path as an NPN Protocol

Division	Factor	NPN Correlate	Functional Rationale
I. Wisdom (*Paññā*)	1. Right View	Foundational *Epistēmē*	The intellectual acceptance that one's conscious desires are not the ultimate authority.
	2. Right Intention	*Prohairesis* for Harmony	A rational commitment to renounce cravings that fuel internal conflict.
II. Ethical Conduct (*Sīla*)	3. Right Speech	Quieting the *Thymos*	Consciously refraining from communication that creates social strife (*Neikos*).
	4. Right Action	*Demiourgos* Framework	Refraining from external harm to create a ceasefire for internal work.
	5. Right Livelihood	*Hormē* Alignment	Ensuring core striving is met through means that promote social cohesion.
III. Mental Discipline (*Samādhi*)	6. Right Effort	*Energeia*	The active application of choice to nurture *Logos*-aligned states.
	7. Right Mindfulness	*Lysis* of *Doxa*	Objective observation that prevents identification with the stories of the self.
	8. Right Concentration	Calming the Rebellion	Unifies the mind, allowing the *Logistikon* to stand down from its futile war.
Cessation (*Nibbāna*)	Liberation	*Dikaiosynē*	The functional state of the *Psyche* operating in alignment with its own deep *Somatic logos*.

I.4 The Mechanism: Surrender to the *Somatic logos* and the End of Strife

The ultimate goal, *Nibbāna*, is understood here not as a supernatural state, but as the **cessation of the internal rebellion**. It is the extinction of the fires of **Internal Misalignment** because the fuel—the ignorant *Doxa* of the conscious mind—has been consumed.[8]

- **The Illusion of the Commander:** The sense of a separate "self" (*attā*) is the headquarters of the rebellion. It is the fiction that is trying to overrule the genetic knowledge of the body. The practice of mindfulness and insight meditation (*vipassanā*) is a direct investigation that reveals this "commander" to be an emergent phenomenon, a temporary confluence of forces, not a solid entity. Seeing this is the final *Lysis*.[9]
- **Alignment and Spontaneous Right Action:** When the rebel commander dissolves, the *Psyche* does not become inert. Instead, it begins to function according to its deepest, most integrated wisdom. Actions arise not from a calculated ego, but from a spontaneous alignment with the situation. This action is inherently ethical because it is no longer generated by the greedy *Orexis* or the defensive *Thymos* of a separate self, but from the integrated intelligence of a system that has made peace with its own nature. This is the embodiment of *Dikaiosynē*.[10]

The "more comfortable existence" is thus the peace that comes when the civil war within the *Psyche* ends. It is the comfort of a system no longer wasting vast resources on an internal conflict between a foolish general and a wise, ancient body. The *Hormē* is no longer frustrated by the *Nous*;

---

8. For the definition of *Nibbāna* as the destruction or "blowing out" of the fires of lust, hatred, and delusion, see *The Connected Discourses of the Buddha*, SN 38.1.

9. For the foundational analysis of the individual as a composite of five aggregates without a permanent self (*Anattā*), see the *Anattalakkhana Sutta* in *The Connected Discourses of the Buddha*, SN 22.59.

10. For the description of the enlightened individual (*Arahant*) who acts without karmic attachment or internal conflict, see *The Dhammapada*, trans. Acharya Buddharakkhita (Kandy: Buddhist Publication Society, 1985), verses 90–99.

APPENDIX I. THE SOMATIC DHAMMA: BUDDHISM AND NPN

instead, the *Nous* becomes a transparent vehicle for the *Hormē's* intelligent and harmonious expression. This is the somatic *Dhamma*—the natural law of a being living in truth with its own history.[11]

I.5 Conclusion: Intuitive Alignment vs. First-Principles Logic

The Somatic *Dhamma* stands as a monumental historical witness to the validity of the NPN framework, discovered not through deductive logic, but through the rigorous empirical observation of the *Psyche* from within. The Buddha and his successors successfully systematized the intuitions of the **Hormē**, proving that internal alignment between the conscious mind and its somatic history leads to a profound cessation of strife.

However, because this tradition lacked the formal **First Principles**—the explicit deductive anchors in the **Archē** and the **Logos**—it remained a system of "insight" rather than a system of "reason." By relying on meditative intuition to navigate, Buddhism was forced to cloak its functional mechanics in the language of mysticism and ascetic renunciation. It demonstrated that alignment works, but it could not explain *why* it works through grounded naturalistic causality.

Where the Buddhist tradition found a harbor of peace through an intuitive surrender to the body's wisdom, NPN provides the actual nautical charts. By moving the discussion from the "mystic feel" of enlightenment to the grounded requirements of **First Principles**, NPN transforms the Buddhist path from a spiritual choice into a biological and logical necessity for any Navigator seeking to sail in truth with reality.

---

11. For the characterization of *Nibbāna* as the "highest happiness" or supreme peace, see *The Dhammapada*, verse 202.

Appendix J

Fortifying the Framework: Core Objections and NPN Rebuttals

"The philosopher is not someone who has a special kind of knowledge; he is someone who no longer takes for granted what others continue to assume without question." – Pierre Hadot[1]

J.1 Introduction: Strengthening the System

A philosophical system is measured not only by the elegance of its positive claims but by its resilience under critical fire. Neo-Pre-Platonic Naturalism, by making strong claims about reality, mind, and morality, invites rigorous scrutiny. This appendix demonstrates that the framework is not merely defensible but is strengthened by the engagement.

A central theme will emerge: **the most potent objections to naturalistic systems are artifacts of a static, synchronic worldview that violate the non-negotiable First Principles of a diachronic reality.** The following rebuttals will not merely counter objections but will demonstrate that they are logically incoherent. For the complete list and justification of all First Principles, see **Appendix B**.

J.2 The Foundation: Objections to Reality and Morality

These objections challenge the very possibility of deriving value from fact and a meaningful morality from a naturalistic foundation.

---

1. Pierre Hadot, *Philosophy as a Way of Life: Spiritual Exercises from Socrates to Foucault*, trans. Michael Chase (Oxford: Blackwell, 1995).

J.2.1 The Epistemological Abyss: NPN's Response to Radical Skepticism

The Objection (The Radical Skeptic's Gambit): "You cannot prove your first principles with absolute certainty. You cannot prove the external world exists, that other minds are conscious, or that your reasoning is reliable. For all you know, you are a brain in a vat, and all your 'models' and 'navigation' are an elaborate illusion. Therefore, your entire system is built on sand."

The NPN Rebuttal: The Pre-emptive Synthesis NPN does not merely respond to this objection; it transcends the very dilemma it presupposes. The radical skeptic is correct in their premise but catastrophically wrong in their conclusion. The error lies in the "spectator theory of knowledge"—the assumption that truth requires a passive, disembodied gaze that mirrors reality without touching it.[2]

NPN grounds what other traditions merely assert: we do not abandon truth for practice; we demonstrate that successful practice is the only valid proof of contact with the *Archē*. To navigate the world without dying is not a "useful fiction"; it is the supreme validation that one's internal map corresponds to the external territory.

1. **NPN Agrees with the Premise:** The framework begins by accepting the skeptic's most powerful point: **Absolute certainty is an illusion.** This is formally established in the Primary Corollary: *The Confidence Gradient of Epistemic Justification*. Certainty is not just unattainable; it is a logical impossibility for a finite mind within the *Archē*.

2. **The Non-Sequitur:** The skeptic's error is the leap from "certainty is impossible" to "therefore, no rational justification is possible." This is a false binary. NPN fills this void not with a weaker form of certainty, but with a stronger form of rationality: the **Confidence**

---

2. For the critique of the "spectator theory of knowledge" as a false standard for certainty, see John Dewey, *The Quest for Certainty* (New York: Minton, Balch & Company, 1929), chap. 1. NPN accepts this critique but rejects the subsequent deflation of truth into mere utility, arguing instead that utility is the *proof* of ontological contact.

APPENDIX J. OBJECTIONS & REBUTTALS

Gradient. This gradient is the only method for making justified decisions with finite data within a dynamic reality.

3. **The Checkmate:** Therefore, the burden of proof is irrevocably reversed. The skeptic must now either:
   - **A)** Prove that absolute certainty is *possible*, thereby disproving the foundational corollary of NPN, or
   - **B)** Present a system for belief and action that is *more functionally adequate* than the Confidence Gradient.

The radical skeptic can offer no competing system, only a state of epistemic paralysis. NPN, by contrast, provides a working, testable, and maximally coherent framework for navigating the very reality whose "proof" the skeptic demands. The skeptic points at the void of the *Apeiron* and says, "You cannot know anything." The NPN Navigator agrees, and replies, "Correct. Now let's talk about what we *can* do. Here is the compass."

Conclusion: The radical skeptic does not refute NPN; they inadvertently demonstrate its necessity. Their position is not a refutation but a pathology—a refusal to use the only rational tools available. By accepting the truth of their premise and building a superior framework from it, NPN doesn't defeat the skeptic. It completes their thought and renders their despair obsolete.

J.2.2 The Is-Ought Problem: The "Master Objection"

The Objection (Hume's Guillotine): "You cannot derive a prescriptive 'Ought' from a descriptive 'Is.' No amount of factual statements about the world can logically lead to a moral conclusion. NPN's entire ethical project is therefore a non-starter."[3]

The NPN Rebuttal: This objection commits a category error by ignoring the **constitutive logic of agency**. It fallaciously applies a rule of propositional logic to the operational logic of a striving system.

---

3. Hume's foundational statement of the Is-Ought problem is found in David Hume, *A Treatise of Human Nature*, ed. David Fate Norton and Mary J. Norton (Oxford: Oxford University Press, 2007), 302 (3.1.1).

1. **It Violates the First Principle of Agency.** The 'Ought' is not a logical deduction from a static 'Is'; it is the operational imperative of a system that *Is* an agent. For a system with *Hormē*, the question "Why ought I to strive?" is biologically and logically incoherent—it is like a fire asking why it ought to be hot. The *Hormē* is the non-negotiable ground of being an agent.
2. **The Derivation is Immanent, Not Logical:**
    - **Is (Fact):** A system with *Hormē* exists.
    - **Is (Functional Truth):** This system operates within a lawful *Archē* (the *Logos*).
    - **Ought (Operational Imperative):** Therefore, to fulfill its constitutive *Hormē*, the system *Ought* to act in accordance with the *Logos*.

The 'Ought' is not invented; it is *discovered* as the necessary condition for the system's own successful existence. Hume's Guillotine only severs 'Ought' from a world of passive facts. It is powerless against a world that contains active, striving agents whose 'Ought' is built into their very 'Is'.[4] The 'Ought' is not derived from an 'Is'; it is the operational output of a constitutive process.

The Humeian Box: The Trap of Modernity

To accept Hume's premise—that "Is" and "Ought" are fundamentally separate categories—is to lose the argument before it begins. This premise trapped the two greatest responses to modernity:

- **Kant (Rationalism)** tried to build a bridge of **Logic** out of the box. He failed because pure reason cannot generate a motive for action without an underlying drive.[5]

---

4. For the seminal argument that 'functional concepts' (such as a watch or a farmer) bridge the fact/value distinction, see Alasdair MacIntyre, *After Virtue*, 3rd ed. (Notre Dame: University of Notre Dame Press, 2007), 57–59.

5. For Kant's attempt to ground morality in pure practical reason, independent of empirical inclination, see Immanuel Kant, *Critique of Practical Reason*, trans. Mary Gregor (Cambridge: Cambridge University Press, 1997), 5:30–35.

APPENDIX J. OBJECTIONS & REBUTTALS

- **Nietzsche (Vitalism)** tried to build a bridge of **Will** out of the box. He accepted that the world was devoid of objective value (Chaos), and concluded that value must be a subjective creation of the strong.[6]

NPN identifies the flaw in the foundation: Reality is not a slide show of static, valueless facts. It is a continuous movie of diachronic process. The "Is" of a living system is not a state of being, but an act of becoming. To see the *Hormē* is to see that the "Ought" is the engine driving the "Is." Value is neither a logical abstraction (Kant) nor a subjective invention (Nietzsche); it is a **functional reality**—the measure of a Navigator's alignment with the *Logos*.

J.2.3 The "Might Makes Right" / Social Darwinist Critique

The Objection: "Your framework simply baptizes evolutionary success as 'moral.' If a stronger, more ruthless or manipulative group exterminates or exploits a weaker, you must deem the victor's traits 'more moral' because they enhanced survival. This is a philosophy of brutality."

The NPN Rebuttal: This objection is a profoundly synchronic misreading that violates the core dynamics of the *Archē*.

1. **It violates the First Principle of Cosmic Dynamics:** The Polarity of *Philia* and *Neikos*. It mistakes a single snapshot of *Neikos* (strife/competition) for a stable, moral state. A society predicated on pure, unchecked *Neikos* is not a stable configuration but a system in a state of perpetual, high-risk expenditure. Its "success" is a temporary illusion that belies its diachronic instability.
2. **It violates the First Principle of Becoming: Diachronic Primacy.** NPN morality is assessed *across time*. A hyper-aggressive strategy invites collective defense from others (*Philia* in response to *Neikos*) and consumes immense internal resources. The framework

---

6. For Nietzsche's argument that values are not objective truths but creations of the will to power, see Friedrich Nietzsche, *Beyond Good and Evil*, trans. Walter Kaufmann (New York: Vintage, 1966), §§ 208–13.

predicts its systemic collapse or transformation. Our moral revulsion is the intuitive recognition of its diachronic maladaptiveness. True moral strength, in the NPN sense, is the **resilient balance** of *Philia* and *Neikos* that ensures long-term stability.

J.2.4 The Charge of Quietism: Justifying the Status Quo

The Objection: "If 'what is, is' a necessary expression of the *Archē*, then isn't the current state of the world—with all its injustice—also necessary? Your system seems to lead to a passive acceptance of the world as it is."

The NPN Rebuttal: This is a fundamental misreading born of a synchronic snapshot that ignores the *engine* of the system.

1. **It violates the First Principle of Becoming: Diachronic Primacy.** *Physis* is not a static state to be accepted, but a dynamic process to be navigated.
2. **It violates the First Principle of Agency: The Primacy of the Hormē.** Our nature as beings with *Hormē* is not passive; it is a driving force for change and improvement across time. The NPN framework provides the objective, diachronic standard (*Dikaiosynē*, *Eudaimonia*) against which the status quo can be judged and actively reformed. It is a tool for the Navigator, not a comfort for the quietist.

J.3 The Bridge: Objections to Mind and Knowledge

These objections challenge the NPN model of the Nous as an emergent navigator and the nature of knowledge itself.

J.3.1 The Hard Problem of Consciousness

The Objection: "You have explained the function of consciousness, but not consciousness itself—the raw feeling of *what it is like*. An unconscious 'philosophical zombie' could, in principle, perform all the same navigational functions. The Hard Problem remains."[7]

---

7. The "Hard Problem of Consciousness" was famously formulated by David Chalmers. See David J. Chalmers, *The Conscious Mind: In Search of a Fundamental Theory* (New York: Oxford University Press, 1996), xi–xii, 4–5.

APPENDIX J. OBJECTIONS & REBUTTALS

The NPN Rebuttal: The so-called "Hard Problem" is a category error generated by the synchronic worldview.

1. **It violates the First Principle of Epistemic Grounding:** The *Somatic logos*. It asks how matter "generates" consciousness as a separate substance in a single moment. But from the diachronic perspective, consciousness is not a separate substance; it is the fundamental **mode of operation** of a sufficiently complex, self-modeling navigator.
2. **It violates the First Principle of the *Nous*: Meta-Cognitive Potential.** The "explanatory gap" is the gap between a third-person, synchronic description of a system and the first-person, diachronic *reality of being* that system. The *Nous* is the *Archē* modeling itself in real-time. The "Hard Problem" is the shock of that model becoming aware of its own existence. The zombie is a logical impossibility because a system of such complexity, evolved for navigation, *is* what consciousness is.

J.3.2 The Problem of Free Will and Determinism

The Objection: "In a deterministic *Archē* governed by the *Logos*, conscious choice is an epiphenomenal illusion. The 'Navigator' is just a passive rider on a predetermined track."

The NPN Rebuttal: This objection searches for a "magic" source of freedom in a synchronic instant, violating the core of the NPN model.

1. **It violates the First Principle of Becoming: Diachronic Primacy.** The power of the *Nous* is not found in breaking physics at a single moment, but in orchestrating behavior *across time*. This is **diachronic causation**. A system that can run high-fidelity simulations of the future and use them to guide its actions is causally powerful in a way a simple stimulus-response machine is not.
2. **It violates the First Principle of the *Nous*: Meta-Cognitive Potential.** The Navigator's freedom is the freedom of **self-correction**. It is the capacity to model its own models, recognize

Doxa, and undergo *Lysis* and *Aporia* to choose a new path. This is not freedom *from* the *Logos*, but freedom *through* the *Logos*.

The ancient dilemma of free will versus determinism is a classic "chicken and egg" problem. The determinist argues that the egg—the prior physical state of the brain—must come first, rigidly determining the chicken of our actions. NPN inverts this causal sequence. We argue that the chicken—the conscious, model-based choice of the Navigator—comes first, and it is this choice that determines the next egg, the new physical state of the world. The Navigator is not a passive passenger in a chain of cause and effect but an active, causal engine that inserts its goals and deliberations into the fabric of reality, thereby laying the eggs of future possibilities that would not have existed otherwise. This view aligns with modern compatibilist accounts that define free will not as a violation of physics, but as the capacity for self-determination through reasoned reflection.[8]

J.3.3 The Charge of Non-Falsifiability and "Just-So" Storytelling

The Objection: "Your 'evolutionary story' about the *Nous* is just-so storytelling. The entire framework is an elegant but non-falsifiable narrative."

The NPN Rebuttal: This objection violates the Primary Corollary: *The Confidence Gradient of Epistemic Justification*.

1. It mistakes the nature of a meta-structure, demanding a level of certainty that is logically impossible for a finite navigator. The framework's strength is its systemic, diachronic coherence and its power to explain a vast range of phenomena without internal contradiction, functioning as a progressive "scientific research programme" in its ability to predict and explain novel facts.[9]

---

8. For a contemporary defense of free will as the capacity for self-determination and reasoned control, which aligns with the NPN concept of the Navigator as a causal engine, see Daniel C. Dennett, *Freedom Evolves* (New York: Viking, 2003), 1–3, 225–27.

9. For the concept of a "scientific research programme" as a series of theories with a coherent, predictive "hard core" and a protective belt of auxiliary hypotheses, see Imre Lakatos, "Falsification and the Methodology of Scientific Research Programmes," in *Criticism and the Growth of Knowledge*, ed. Imre Lakatos and Alan Musgrave (Cambridge: Cambridge University Press, 1970), 91–196.

2. It is, in fact, **falsifiable**. The framework would be invalidated by empirical evidence that fundamentally contradicts its predictions—for example, a society thriving for centuries on principles of pure, unchecked *Neikos*, or the discovery of a fundamental force that is neither *Philia* nor *Neikos* (violating *The First Principle of Cosmic Dynamics*). It operates on a confidence gradient; it is the best model available that does not self-destruct upon its own logical premises.[10]

J.4 The Boundary: Metaphysical and Theological Objections

These are the deepest objections, challenging the naturalistic worldview itself.

J.4.1 The Theological Objection: The "Death of Meaning"

The Objection: "An impersonal, mechanistic *Archē* cannot be the source of ultimate meaning or 'The Good.' You cannot bootstrap moral value from this barren foundation. Without a transcendent source, life is meaningless."

The NPN Rebuttal: This objection presumes meaning must be transcendent, thereby violating the First Principle of Agency: *The Primacy of the Hormē*.

1. **NPN offers immanent, diachronic meaning.** Purpose emerges *within* the system for subsystems that have constitutive goals across time. The feeling of meaning is the subjective correlate of successful diachronic alignment between a Navigator's actions, models, and the *Logos* of the *Archē*.
2. **It violates the First Principle of Knowledge: The Impotence Before the *Apeiron*.** To posit a transcendent source of meaning is to make a claim about the *Apeiron*—a domain for which no empirical grounding is possible. It is an attempt to generate knowledge from an information vacuum.

---

10. For Popper's core thesis on falsification, see Karl Popper, *The Logic of Scientific Discovery* (London: Routledge, 1959), chap. 1.

J.4.2 The Evolutionary Mismatch & Spandrel Argument

The Objection: "Consciousness and our moral faculties might be non-adaptive byproducts—'spandrels.' The 'inner movie' and our sense of justice may have no functional utility."

The NPN Rebuttal: This is spectacularly un-parsimonious and relies on a synchronic view.

1. **It violates the First Principle of Becoming: Diachronic Primacy.** It ignores that natural selection is a diachronic process that relentlessly prunes non-functional, energy-intensive traits. The *Nous* is the most complex and metabolically costly system in the body. To propose it is a mere accident is to ignore the mountain of evidence showing its exquisite design for solving the problem of navigation across time.

2. **It violates the First Principle of Emergent Complexity: The Potential for Nous.** The *Archē* is the kind of system that *generates* navigators. To claim its most sophisticated navigational instrument is a functionless byproduct is to profoundly misunderstand the generative nature of the cosmos.[11]

J.4.3 The Problem of Other Minds & Solipsism

The Objection: "You cannot prove other minds exist. Your entire moral framework, which depends on social beings, is built on sand."

The NPN Rebuttal: This objection is a specific instance of the Radical Skeptic's Gambit, and it fails for the same fundamental reasons. It is a synchronic demand for impossible certainty that violates the Primary Corollary: *The Confidence Gradient of Epistemic Justification*.

---

11. For the concept of "spandrels," see S. J. Gould and R. C. Lewontin, "The Spandrels of San Marco and the Panglossian Paradigm: A Critique of the Adaptationist Programme," *Proceedings of the Royal Society of London B* 205 (1979): 581–98.

APPENDIX J. OBJECTIONS & REBUTTALS

As with the existence of the external world, we do not have certainty of other minds, but we possess the highest possible **confidence**. The inference to other minds is the only model that provides a coherent, non-contradictory, and functionally effective explanation for the behavior we observe. It is a hypothesis that is continuously and overwhelmingly validated by its predictive success in every social interaction. To reject this model in favor of solipsism is to embrace a functionally sterile and logically self-undermining position, as outlined in our rebuttal to **Radical Skepticism (see previous section)**.

J.5 Conclusion: The Resilience of the Diachronic Navigator

The objections leveled against Neo-Pre-Platonic Naturalism are profound, yet they are consistently products of a **synchronic worldview** that analyzes reality in frozen snapshots. As demonstrated, these critiques invariably **violate one or more of the system's non-negotiable First Principles**.

The NPN framework, grounded in the **Diachronic Turn**, reveals these objections not as fatal blows, but as misapprehensions stemming from a flawed perspective on the nature of reality, mind, and time. A philosophy that cannot answer its critics is a dogma. A philosophy that is strengthened by them, and in turn demonstrates that their objections stem from a failure to grasp its foundational principles, is a robust and powerful tool for navigation. The Navigator does not seek a harbor of certain truth, but masters the art of sailing the boundless, dynamic, and lawful *Archē*.

Appendix K

First Philosophy: The Boundary Condition: A Geometric Derivation of Reality from the Logic of Distinction

"Draw a distinction." - George Spencer-Brown

Introduction: The Empty Starting Point

Any claim about what exists must first account for how one thing can be distinguished from another. Traditional philosophy begins with something—water, mind, substance, God—and then attempts to explain distinction. This paper begins with distinction itself.

We start with a blank space. The first act is to draw a line, creating an inside and an outside.[1] This act requires no prior metaphysics. It is the precondition for any metaphysics at all.

From this single starting point—the logic of the boundary—we will derive the fundamental structures of reality: the necessity of an indeterminate ground, the possibility of plurality and change, the limits of knowledge, and the origin of value. The argument proceeds as a geometric proof. No appeals to experience, intuition, or empirical fact are made, save one: that

---

1. The operation of drawing a boundary to create existence is formally isomorphic to the Calculus of Indications developed by George Spencer-Brown, which begins with the single injunction: "Draw a distinction." See George Spencer-Brown, *Laws of Form* (London: George Allen and Unwin, 1969). Spencer-Brown demonstrated that the laws of logic and arithmetic emerge inevitably from the primary act of severing a space, a mathematical validation of the metaphysical separation (Adikia) described here.

a distinction can be drawn. If you understand the difference between this word and the next, you have performed the act.

This is First Philosophy: the study of what is first in the order of being. Not first in time, but first in necessity. We do not claim to have discovered a new entity, but to have uncovered the **geometric necessity that every entity must satisfy**. The Boundary Condition is not a part of the world; it is the shape of worldhood.[2]

Part I: The Geometry of Distinction

The following theorems describe the structural preconditions of worldhood. They apply equally to a universe of atoms, a universe of minds, or a universe of pure forms. Before we can discuss the nature of the bounded, we must establish the logic of the boundary.

1. The First Move: The Boundary

Consider any entity. To be *that* entity and not another, it must have a limit. This limit is not an added feature; it is the condition of being an individual. We call this limit a **boundary**.

Definition 1.1 (Boundary): A closed distinction that separates an interior from an exterior.[3]

A boundary is "closed" because if it had gaps, the interior would merge with the exterior and the distinction would fail. A circle with a break is not a circle.

---

2. We reclaim the term "First Philosophy" in the Aristotelian sense—as the study of "being qua being." See Aristotle, *Metaphysics*, in *The Complete Works of Aristotle: The Revised Oxford Translation*, ed. Jonathan Barnes (Princeton, NJ: Princeton University Press, 1984), .1, 1003a21. However, whereas Aristotle sought the primary substance (*ousia*), this inquiry locates the "first" not in a substance, but in the geometric boundary condition required for any substance to exist.

3. This definition aligns with the geometric axiom established by Euclid, who defined a boundary not as an object, but as a limit: "A boundary is that which is an extremity of anything." Euclid, *Elements*, Book I, Def. 13. Metaphysically, this establishes that a boundary is not a constituent part of a substance, but the condition of its finitude.

First Philosophy

Observation 1.2: To be a determinate entity is to be bounded. Determinacy implies finitude; finitude implies a limit.

2. The Second Move: The Interstitial Necessity

Consider two bounded entities, A and B. For them to be distinct, their boundaries must not overlap or merge. If their boundaries touched perfectly, they would share a boundary.

Theorem 2.1 (Interstitial Necessity): For any two distinct bounded entities A and B, there exists a region R such that:

1. R is not part of A.
2. R is not part of B.
3. R lies strictly between the boundary of A and the boundary of B.[4]

Proof: Suppose no such region R exists. Then the boundaries of A and B are in direct contact.

When two boundaries are in direct contact, they form a **single continuous boundary** enclosing both A and B together. The line between them is not two boundaries but one shared division.

Consider two islands in an ocean. If the islands touch, they become one island with two hills. For them to be two distinct islands, there must be water between them. That water is not part of either island. It is the interstitial space that makes their separateness geometrically real, not just nominal.

Therefore, without an interstitial region R, A and B are not two distinct bounded entities. They are subdivisions of a single bounded whole. Their distinction is nominal, not geometric.

---

[4]. The necessity of the "empty" or interstitial for the existence of the "full" (determinate) was first articulated in the Dao De Jing: "We mold clay into a vessel; but it is on the empty space within that the use of the vessel depends." Laozi, *Tao Te Ching*, Ch. 11. Just as the utility of the wheel depends on the empty hub, the distinctness of any bounded entity depends geometrically on the indeterminate region that is not the entity.

Thus, for A and B to be geometrically distinct—each with its own closed, unshared boundary—a region R must exist between them. ∎

3. The Third Move: The Nature of the Interstitial

What is R? If we claim R is itself a bounded entity C, then by Theorem 2.1, new interstitial regions are required between A and C, and between C and B. This either continues infinitely or terminates in a region that is **not bounded**.

Theorem 3.1 (Indeterminacy of the Interstitial): The interstitial region R cannot be a bounded entity. It must be *indeterminate* relative to A and B.[5]

Proof: Assume R is a bounded entity D. Then by Theorem 2.1, there must be interstitial regions R_1 between A and D, and R_2 between D and B. If R_1 and R_2 are also claimed to be bounded, the regress continues without end. An infinite regress of bounded entities provides no foundation for distinction. The only coherent stopping point is an interstitial region that is **not itself bounded**. Therefore, R must be indeterminate. ∎

"Indeterminate" here means: lacking a closed boundary of its own. It is not another figure, but the **ground against which figures appear**.[6]

---

5. This derivation provides the geometric proof for Spinoza's dictum *determinatio est negatio* ("determination is negation"). See Benedict de Spinoza, "Letter 50," in *Spinoza: Complete Works*, ed. Michael L. Morgan, trans. Samuel Shirley (Indianapolis, IN: Hackett Publishing Company, 2002). Spinoza recognized that to define a thing is to limit it, which necessarily implies a negation of the infinite background. Here, we prove that this background cannot itself be another determinate thing without generating an infinite regress, confirming that the ground of existence must be the indeterminate Apeiron.

6. This theorem formalizes Anaximander's insight that the origin of determinate things cannot itself be determinate (elemental), but must be the *Apeiron* (Boundless). For a comprehensive derivation of the *Apeiron* as a relational indeterminacy rather than a material substance—and a refutation of the standard Aristotelian interpretation—see Eli Adam Deutscher, "Anaximander and the Zero Principle: The Relational Ontology of the Apeiron" (Preprint, 2026), https://neopreplatonic.com/papers/anaximander/. Deutscher demonstrates through etymological analysis that Anaximander's terminology (*a-peiron*, *a-dikia*) describes a logic of boundary dependent on an indeterminate complement, anticipating the Zero Principle.

Corollary 3.2 (The Boundary Condition): All bounded entities require an indeterminate interstitial ground.

This is not an assumption. It is the **geometric necessity** revealed by the logic of distinction.

To visualize the necessity of the indeterminate ground (Theorem 3.1 and Corollary 3.2), we may perform the following thought experiment.

Construction: Consider an unbounded **continuum** (a 2D plane). Upon it, we draw two concentric circles:

1. **A small inner circle (A):** Representing a determinate entity or system.
2. **A larger outer circle (U):** Labeled "The Totality" or "The Universal Container," intended to bound all possible existence.

The Geometric Error: The construction of U creates an immediate topological contradiction regarding the nature of the ground. By Definition 1.1, the boundary of U must separate an interior from an exterior. If U has an exterior, it is not "The Totality"; it is merely a larger determinate object (U') existing within a wider field. If we attempt to resolve this by drawing a third circle (U'') to encompass the exterior of U', we initiate an infinite regress of containment. We are chasing the ground by drawing larger figures, but every figure presupposes the ground upon which it is drawn.

The Topological Solution: The ground of A is not the larger circle U. The ground is the **plane itself** upon which both A and U are inscribed.

- The circles are **Determinate Figures** (defined by limits).
- The plane is the **Indeterminate Field** (the condition for limits).
- **You cannot draw the plane on the plane itself.** To draw the plane would require drawing a boundary *around* it, which would turn the plane into a figure (U) and require a new background plane to sustain it.

The Insight: Now that we have finished the thought experiment, think of what made it possible at all: *"Consider an unbounded continuum."* To even conduct the experiment, we had to stipulate the ground. That is not an accident; that is the precondition for any distinction at all.

Part II: From Geometry to Cosmos

Having established the Boundary Condition—that bounded entities require an indeterminate interstitial ground—we now explore its consequences. What kind of world does this geometry demand? We will derive the fundamental features of any possible reality that contains distinct things: plurality, dynamism, and the limits of knowledge. Each follows directly from the logic of boundaries, without speculation about the nature of the ground.

4. Deriving Plurality: Why There Cannot Be Only One

Could a reality contain exactly one bounded entity and nothing else? Let us examine the geometry of this proposition.

Theorem 4.1 (The Impossibility of Singular Bounded Totality): A reality consisting of a single bounded entity A, with an exterior of absolute non-existence, is geometrically incoherent.

Proof: Suppose A is the only existent, bounded entity. Its exterior is defined as absolute non-existence.

Recall that a boundary, by Definition 1.1, is a *distinction* that separates an interior from an exterior. A distinction requires a difference. For the exterior to serve as a term in this distinction, it must be a *contrast*.

For entity A to be bounded, its exterior must be a geometric term capable of contrast—an "other" to its "self." This other cannot be absolute non-existence, as nothingness provides no surface for a boundary to abut against. The exterior must therefore be a positive, albeit indeterminate, ground. If one were to deny this and claim A exists in a void of non-existence, the concept of A's boundary loses all meaning. Without a contrasting ground, the distinction "inside/outside" vanishes, and A effectively expands to fill all conceptual space.[7]

Therefore, a boundary cannot have absolute non-existence as one of its sides. The exterior must be a *positive term* in the geometry of distinction. It need not be another bounded entity, but it must be *something* that can stand in contrast to the interior. By Theorem 3.1, we have already identified this positive term: the **indeterminate interstitial ground**.

Thus, a solitary bounded entity A is not alone. It exists in necessary relation to the indeterminate ground. The ground is not A, but it is not nothing—it is the logically necessary exterior term, the "not-A."

The configuration is not "A and nothing," but "A and the Ground." The ground, being indeterminate, does not possess the bounded unity of a "One." It is the field within which A appears. This field, as the domain of "not-A," is the space where the geometric operation of bounding can, in principle, be repeated. ∎

Corollary 4.2: The existence of one bounded entity logically implies the **possibility of others**. The ground is the space of this possibility.

5. Deriving Dynamism: Why There Must Be Change

The necessity of an interstitial ground does not, by itself, guarantee change. Could a reality of bounded entities be static? We now prove that true stasis—the absolute prevention of interaction—is geometrically incoherent. The possibility of interaction is not an added feature but a necessary consequence of the geometry we have established.

---

[7]. The central polemic of Parmenides is precisely this demonstration: he recognized that the Anaximandrian *Apeiron*—the boundless, indeterminate ground necessary for determination—cannot be made an object of determinate thought or speech without logical contradiction. By banning "What Is Not" from discourse, he performed a *reductio ad absurdum* that exposes this crisis, showing that logic alone, stripped of its indeterminate background, collapses into a frozen "One." See Eli Adam Deutscher, "Parmenides the Polemicist: The Eleatic Crisis and the Indeterminate Ground of Thought" (Preprint, 2026), https://www.neopreplatonic.com/papers/Parmenides/.

Theorem 5.1 (The Causal Neutrality of the Interstitial Ground): The interstitial ground R between bounded entities A and B is **causally neutral**. It possesses no determinate properties that govern interaction.

Proof (By contradiction of a determinate function): Assume R is **not** causally neutral. For it to be non-neutral, it must have a specific causal disposition—it must *do* something. It could act as an absolute barrier, a selective filter, or a mediating transformer of influence between A and B.

Any such specific function is a **determinate property**. A region with a determinate causal property is not a passive background; it is a **causal agent** with a defined role and identity within the system.

However, by **Theorem 3.1 (Indeterminacy of the Interstitial)**, R is indeterminate. It cannot have a determinate identity or properties. Therefore, our initial assumption leads to a contradiction: R cannot simultaneously be indeterminate *and* possess the determinate property of a causal function.

Thus, R cannot have any causal function. The only consistent alternative is that R is **causally inert**. It does not act, prohibit, or permit. It is the geometric medium, not a participant. ∎

Corollary 5.2 (Interaction is Not Geometrically Precluded): Since the ground is causally neutral, **no geometric law inherent to the foundational structure of reality forbids interaction** between bounded entities.

This is a pivotal shift. It means that within a reality of distinct bounded entities (A, B, C...), the question of whether they interact is **not settled by the nature of the ground**. The ground does not *cause* interaction, but its neutrality means there is **no first-principle veto** against it. Interaction becomes a **contingent possibility** dependent on the properties and configurations of the bounded entities themselves.

Corollary 5.3 (The Primacy of Process): In a reality with multiple bounded entities capable of being affected, **stasis is the *geometrically***

special case, while dynamism is the *geometrically general* possibility.

If entities have properties that allow them to influence one another (e.g., mass, charge, semantic content), and nothing in the fundamental geometry forbids it, then interaction—and thus change—is a **structurally enabled possibility**. A static universe requires *every* entity to be inert or for *all* configurations to preclude interaction—a highly specific, restrictive set of conditions. The geometry of boundaries and a neutral ground establishes a cosmos where the **potential for process is inherent**. Being (a stable, bounded pattern) is a temporary, sustained equilibrium within the broader, fundamental *possibility* of Becoming.

6. Deriving the Limits of Knowledge

Knowledge is a relation between a knower (a bounded entity) and that which is known. What can be known within the Boundary Condition?

Axiom 6.1 (Modeling): To know something is to generate and maintain a bounded internal structure (a model) whose relations correspond to the relations of the thing known.

Theorem 6.2 (The Domain of the Knowable): Only bounded entities can be modeled, and thus directly known.

Proof: A model is itself a bounded entity—a set of related distinctions within the knower. By Definition 1.1, it has an interior (the model's content) and an exterior (what the model excludes).

To model something is to map its boundary and internal structure. The target must therefore *have* a boundary and internal structure to map.

The indeterminate interstitial ground, by Theorem 3.1, has no boundary. It possesses no interior/exterior structure. Therefore, there is no geometric structure to which a bounded model can correspond. Any attempt to create a "model of the ground" would, by necessity, impose a boundary upon it, creating a bounded representation *of* it, not a correspondence *to* it. ∎

Corollary 6.3 (The Unknowable Ground): The indeterminate ground is necessarily unknowable in principle. This is not a contingent limitation of any particular cognitive system, but a geometric limit of the modeling relation itself.

Important Clarification: To construct a bounded model of the ground is not to "falsify" the ground—the ground is indifferent. The error is in the knower. The model is *Doxa*—unjustified belief: a bounded construct mistaken for a map of the boundless. It is a category error, producing the illusion of understanding.

Theorem 6.4 (Navigational Knowledge): Knowledge is the successful modeling of **bounded entities and the patterns of their interaction**. The ground, being causally neutral and indeterminate, is not a term in these models. Navigation is the application of such models to anticipate and respond to the behavior of other bounded entities. The ground is not an object of navigation; it is the **condition for the possibility of navigation**—the open space through which movement and influence occur.

Part III: First Philosophy as Navigation

The Boundary Condition is not merely a description of reality—it is an **operating principle**. To be a bounded entity in a world of bounds and interstitial ground is to face a fundamental task: **navigation**. In this final part, we show how the core domains of human concern—meaning, knowledge, and ethics—emerge not as social conventions or subjective preferences, but as necessary forms of this navigation. They are the logic of a bounded entity persisting through time within the geometric reality we have derived.

7. Semantics: Meaning as Boundary-Drawing

A concept is a cognitive boundary. Its meaning is not a list of intrinsic properties, but its **position within a field of contrast**.

Axiom 7.1 (Concept as Bound): A concept C is a bounded distinction drawn within cognitive space. Its interior is the set of referents or instances it includes; its exterior is what it excludes.

Theorem 7.2 (The Semantic Ground): The meaning of a concept C depends on the indeterminate conceptual ground against which it is distinguished.

Proof: By the Boundary Condition (Corollary 3.2), the bounded entity C requires an indeterminate ground. In the cognitive domain, this ground is the **space of potential meaning**—the unstructured field of possible distinctions and experiences not yet carved into concepts.

If one attempts to define C solely by its relation to other bounded concepts (D, E, F...), one encounters the **dictionary regress**: each defining concept itself requires definition, leading either to circularity or infinite regress.

The regress stops when concepts are anchored in their **functional correspondence to the determinate ground of objective, structured reality (*Archē*)**—that is, in their successful use for navigating the world of bounded entities and patterns. Therefore, a concept acquires determinate meaning only when it is anchored in a successful prediction of interaction. The "meaning" is not what the object is in isolation, but what it does in relation to the agent's striving (Hormē). ■

Corollary 7.3 (Semantic Grounding in the *Archē*): The semantic regress stops in successful navigation: a concept earns its content through its use in coordinating action within the structured world. The meaning of "Chair," therefore, is not "a piece of furniture distinct from a table." The meaning of "Chair" is "that-which-supports-the-body-against-gravity." It is defined by its functional purpose within the *Archē*. The definition is anchored in the Somatic: the agent feels the fatigue of gravity and discovers a rigid pattern in the *Archē* that counteracts it. This functional correspondence is the "absolute" at the bottom of the semantic chain. ■

8. Epistemology: Knowledge as Map-Making

To know is to possess a model that corresponds to reality. But what can a bounded knower model?

Axiom 8.1 (Model as Bound): A cognitive model M is a bounded, internal structure within the knower.

Theorem 8.2 (The Map-Territory Relation): A model M can only correspond to aspects of reality that are themselves bounded.

Proof: By Theorem 6.2, only bounded entities can be modeled. A model maps boundaries to boundaries. The territory it maps must therefore have discernible, bounded features—objects, patterns, regularities.

The indeterminate ground, having no boundary, cannot be modeled. Any "model of the ground" would impose a false boundary, creating *Doxa* (unjustified belief) rather than correspondence. ■

Corollary 8.3 (Navigational Knowledge): Knowledge is therefore **navigational**. It is a map of the bounded landmarks and the stable relations between them within the indeterminate field. We can know the routes, the distances, the reliable patterns—the *logos* of how bounded entities behave in the ground. We cannot know the ground itself, but we can learn to move through it effectively.

Theorem 8.4 (The Confidence Gradient): The truth of a model is not a binary state but a gradient of **functional alignment**. A model is "true" insofar as it reliably guides the knower through the territory without collision or failure. Confidence increases with successful navigation and decreases with predictive error.

9. Ethics: Value as Boundary Maintenance

An agent is a special kind of bounded entity. A passive bounded entity (like a rock) has a boundary, but it possesses no internal mechanism to regulate it; its persistence is entirely contingent on the neutrality of the environment. If the flow of events dictates its erosion, it erodes. An active bounded entity (an Agent) is one that introduces a bias into this flow.

Definition 9.1 (Agent): A bounded entity that possesses internal mechanisms to **alter the predetermined flow** of interaction with the environment. It acts to redirect causal trajectories that would otherwise lead to dissolution, thereby maintaining its boundary and internal organization across time. This creates a geometric distinction between the *Passive Boundary* (which endures only until disrupted) and the *Active Boundary* (which resists disruption).

Theorem 9.2 (The Necessity of Causal Bias): For an Active Boundary to persist within a neutral ground that permits interaction (see Theorem 5.2), it must be constitutively oriented toward boundary maintenance. This is formalized as T7: The Entropic Asymmetry in Appendix B

- **The Inertial State:** Because the ground is causally neutral (Theorem 5.1), the "predetermined flow" of any bounded system interacting with a dynamic environment is towards entropic equilibrium (loss of distinction).
- **The Agentic Act:** Therefore, persistence requires work—a specific deviation from the inertial path. The Agent must introduce a *Causal Bias* into the system to favor continuity over dissolution.
- **Conclusion:** This capacity to change the flow is the geometric definition of **Will** and the structural definition of **Life**. *Note: This geometric necessity—that persistence requires the active deflection of inertia—is the formal derivation of* **First Principle 6 (Primacy of the Hormē)** *in the main text.*

Theorem 9.3 (The Origin of Value): "Good" and "bad" emerge as functional evaluations made by an agent's navigational system regarding interactions that are causally relevant to its boundary maintenance.

Proof: For an agent A, any interaction with the environment (other bounded entities) either supports or frustrates its boundary maintenance.

- An interaction that supports or enhances A's bounded integrity is *good* for A.

- An interaction that degrades or threatens A's bounded integrity is *bad* for A.

This evaluation is not a subjective whim; it is a geometric fact about the relationship between the agent's structure and the causal consequences of interaction. If an action leads to the dissolution of the agent's boundary, that action is objectively bad for that agent, as it negates the condition for the agent's existence. ∎

Corollary 9.4 The Meta-Cognitive Condition for Navigation

The capacity for navigation described in Theorem 9.4 and the evaluative function described in Theorem 9.3 require a specific cognitive capability: the system must be able to **model not only its environment but also its own modeling process**. This is the *Meta-Cognitive Potential* of the *Nous* (FP9).

An agent that merely reacts to stimuli is not navigating; it is being buffeted. True navigation requires the agent to:

1. **Maintain a self-model:** Identify itself as the bounded entity whose persistence is at stake.
2. **Monitor for misalignment:** Detect discrepancies (errors) between its internal model and the state of the *Archē*, and between its actions and the requirements of its *Hormē*.
3. **Execute correction:** Adjust its model or behavior to reduce these errors, thereby **increasing the functional confidence** and **decreasing the misalignment** in both the epistemic and practical domains.

This self-referential operation is the **engine of agency**. It is what transforms a bounded system into a navigator. The capacity for this operation is therefore a structural prerequisite for the ethical and epistemic functions derived in this section. The "Ought" becomes operational for a system that can evaluate whether its current "Is" is aligned with its own continued existence.

Corollary 9.5 (The Is-Ought Collapse): For an agent, the "ought" is directly implied by the "is." The fact that an agent *is* a bounded organization striving to persist (*Hormē*) implies that it *ought* to act in ways that fulfill that striving. The question "Why ought I to preserve myself?" is incoherent—it asks for a reason outside the very condition of being an "I."

Theorem 9.6 (Ethical Isomorphism): Ethical error (vice) and epistemic error (falsehood) are isomorphic. Both are states of **misalignment** between the agent's internal model/action and the *logos* of the bounded environment.

- **Epistemic error:** A model that mispredicts the behavior of bounded entities leads to navigational failure.
- **Ethical error:** An action that ignores the geometric constraints of boundary maintenance leads to organizational failure.

Both reduce to the same geometric flaw: a mismatch between the agent's internal boundaries and the external boundaries it must navigate.

Conclusion: The Boundary Condition

We began with a blank space and the act of drawing a distinction. From this alone, we have derived the fundamental architecture of any possible reality containing distinct entities:

1. **The Geometry:** Bounded entities require an indeterminate interstitial ground (The Boundary Condition).
2. **The Cosmology:** This implies a dynamic, pluralistic field where interaction and change are necessary possibilities.
3. **The Epistemology:** Knowledge is limited to the bounded and takes the form of navigational map-making.
4. **The Ethics:** Value originates in the geometric imperative of boundary maintenance for persistent agents.

This is First Philosophy. It is first not because it is historically prior, but because it is **foundational in the order of necessity**. Every other inquiry—physics, biology, psychology, sociology—presupposes a world of distinct, interacting entities. We have shown what such a world must look like, at the barest geometric level.

The implications are profound:

- **The end of substance metaphysics.** The fundamental question shifts from "What is everything made of?" to "How are boundaries drawn and maintained?"
- **A resolution of the fact-value gap.** Value is revealed as the functional logic of persistence for any bounded agent.
- **A clear epistemic limit.** We obtain a principled humility about what can be known, derived from geometry, not skepticism.
- **A universal navigational imperative.** Life, mind, and society can be understood as layered, complex forms of boundary navigation in an indeterminate field.

The task of philosophy—and indeed, of any rational agent—is therefore not to seek a disembodied "view from nowhere," but to **improve the accuracy of its maps and the wisdom of its navigation**. To think well is to draw boundaries that correspond to the world's own distinctions. To act well is to move in ways that honor the geometric conditions of one's own existence and the existence of others.

We have reached the foundation. It is not rock, but **relation**: the relation of boundary to ground. Everything that is, is bounded. Everything that is bounded, exists in necessary relation to the boundless. To understand this is to begin to see the world as it truly is: a dynamic, articulate, navigable whole.

Ode to the Giants

"With these great Greek thinkers we stand at the boundary of the true and proper philosophy of the Greeks. All subsequent thinkers... appear as a decline in comparison with this great beginning... Here the Greek genius philosophizes for the first time without a veil, naked and daring." – Friedrich Nietzsche[8]

From Fragments to Framework

We all stand on the shoulders of giants. The goal of Neo-Pre-Platonic Naturalism (NPN) is not to prove the giants of philosophical history wrong; it is to prove them fundamentally, profoundly **right**. The greatest thinkers—from Thales to Aristotle—did not deal in convenient fictions; they were navigating the real contours of the *Archē* with instruments of unparalleled genius.

The fundamental conflict that has defined Western philosophy for over two millennia is not a debate over truth, but a structural incoherence caused by the Synchronic Flattening. Because the giants were forced to map a four-dimensional process using three-dimensional static models, their magnificent discoveries inevitably led to paradox and schism—to be right in one dimension often meant being catastrophically wrong in another.

This appendix serves as a final, systematic **Homage** to their legacy. It demonstrates that every core concept discovered by the philosophical giants was a necessary, real landmark of the cosmos. The resulting NPN framework is not a philosophical patchwork, but the definitive synthesis—the first complete map that coherently integrates their partial views. The

---

8. Friedrich Nietzsche, *The Pre-Platonic Philosophers*, trans. Greg Whitlock (Urbana: University of Illinois Press, 2001), 31–32.

fact that a system logically deduced from the First Principles of a striving agent (*Hormē*) within a lawful reality (*Logos*) aligns so precisely with their fragments is the deepest, most humbling form of validation possible. We do not critique their vision; we finally give them the voice to tell their piece of the puzzle.

Thales established the foundational imperative of philosophy itself: the search for a single, unified source of all things—the *Archē*. While his candidate (water) was provisional, his true legacy was the methodological turn from myth to a rational account of a coherent physical reality.[9]

"The world is one, and it can be known by reason."

Anaximander advanced this project by identifying the *Apeiron*—the boundless, indefinite ground from which all determinate things emerge and to which they return according to a necessary order he called *Dikē*. In NPN, the *Apeiron* is the absolute boundary of the knowable, and *Dikē* is the causal, impersonal justice of the *Logos*.[10]

"Every map is smaller than the territory it represents."

Heraclitus correctly identified the fundamental character of reality as perpetual flux, governed by an immanent, divine *Logos*. His insight establishes the principle of Diachronic Primacy within NPN: reality is a lawful process, not a static picture, and the *Logos* is its constitutive structure.[11]

"Existence is not a noun, but a perpetual verb."

Xenophanes established the constitutive epistemic condition of the finite *Nous*: that mortals, through "long seeking" (*zētountes*), discover what

---

9. For Thales' claim that the earth rests on water as the fundamental principle, see G. S. Kirk, J. E. Raven, and M. Schofield, *The Presocratic Philosophers: A Critical History with a Selection of Texts*, 2nd ed. (Cambridge: Cambridge University Press, 1983), 84.

10. For Anaximander's concept of the *Apeiron* as the fundamental principle, see Kirk, Raven, and Schofield, *The Presocratic Philosophers*, KRS 101B. For the doctrine of "cosmic justice," see KRS 110.

11. For the river fragments and the Logos as the universal principle, see Heraclitus in Kirk, Raven, and Schofield, *The Presocratic Philosophers*, KRS 214 and 194–96.

is better (*ameinon*), but can never grasp clear truth (*to saphes*) with certainty.¹²

> "*Certainty is the illusion; seeking is the reality.*"

The Atomists (Leucippus & Democritus) provided the material blueprint of the cosmos with their theory of atoms and void—the "Full and the Empty." NPN recognizes this as an accurate, if incomplete, map of the *Hylē* (the hardware of reality).¹³

> "*We mapped the hardware of the cosmos: the Full and the Empty.*"

Parmenides, through pure logic, discovered the unity, permanence, and self-identity of "What Is." NPN synthesizes this not as a description of the cosmos, but as the discovery of the internal architecture of the *Nous*—the stable, unified structure of subjective consciousness itself.¹⁴

> "*I discovered the unchanging logic, not of the cosmos, but of the self that observes it.*"

Empedocles identified the two fundamental, impersonal cosmic forces: *Philia* (Love/Attraction) and *Neikos* (Strife/Repulsion). In NPN, this is the Exhaustive Polarity—the dual engine of all Emergent Complexity, driving the cohesive and selective processes of cosmic and biological evolution.¹⁵

> "*All creation is but the eternal, necessary tension of attraction and strife.*"

Socrates turned philosophy inward, acting as a physician of the *Psyche*. His method of *Elenchus* induced *Aporia*—a productive void—through the

---

12. Xenophanes, DK 21 B34 and B18; see James H. Lesher, *Xenophanes of Colophon: Fragments* (Toronto: University of Toronto Press, 1992), 150–69.
13. For the atomic theory, see Kirk, Raven, and Schofield, *The Presocratic Philosophers*, KRS 549.
14. For Parmenides' doctrine of "What Is," see Kirk, Raven, and Schofield, *The Presocratic Philosophers*, KRS 296.
15. For Empedocles' four roots and the two forces, see Kirk, Raven, and Schofield, *The Presocratic Philosophers*, KRS 346 (roots) and 360 (forces).

Lysis (dissolution) of unexamined belief (*Doxa*). NPN formalizes this as the essential therapeutic and cognitive protocol for the Navigator.[16]

> *"The greatest journey begins by proving your own map is false."*

Plato (Middle Period) provided one of the most enduringly accurate maps of the human *Psyche*, correctly identifying its stratified nature—the interplay of appetitive, spirited, and rational parts (*Orexis, Thymos, Logistikon*).[17] His genius was to frame its proper state as a harmonious governance (*Dikaiosyne*), a profound insight NPN adopts directly.[18]

Plato (Late Period) performed a profound Diachronic Reckoning, systematically deconstructing his own earlier system and conceding that cosmological accounts are merely "likely stories." This anticipates the NPN epistemology of the Confidence Gradient and his political turn toward a "second-best" state grounded in immanent law.[19]

> *"Only the likely story remains, for certainty belongs not to the mortal mind."*

Aristotle provided the definitive synchronic account of reality's two-part nature, perfecting the concept of hylomorphism—that every being is a compound of dynamic potentiality (*Hylē*) and structuring principle (*Form*).[20] He correctly identified *Eudaimonia* as the *Telos* of human life, grounding it in the excellent activity of the human function.[21]

> *"I mapped what is, perfecting the language of substance and structure."*

---

16. For Socrates' description of his mission, see Plato, *Apology*, 29d–31c. For the induction of aporia, see Plato, *Meno*, 80a–b.
17. For the tripartite *Psyche*, see Plato, *Republic*, 439d–441c.
18. For the Theory of Forms as a transcendent reality, see especially Plato, *Phaedo*, 65d–66a.
19. For the "likely story," see Plato, *Timaeus*, 29c–d. For the "second-best" state, see Plato, *Laws*, 875a–d.
20. For the doctrine of hylomorphism, see Aristotle, *Physics*, 194b16–195b30.
21. For *Eudaimonia* as the highest good, see Aristotle, *Nicomachean Ethics*, 1098a7–18. For the "function argument" (*ergon*) which defines the human good as activity of the soul in accordance with excellence, see *Nicomachean Ethics*, 1097b22–1098a20.

Glossary of Key Terms

"The beginning of wisdom is the definition of terms."
– Socrates[22]

"It is the mark of an educated man to look for precision in each class of things just so far as the nature of the subject admits."
– Aristotle[23]

A Note on the Lexicon: The terms below are defined according to their specific technical meaning within the NPN meta-structure. They are building blocks of the system, and their reclaimed Greek forms are essential for maintaining conceptual precision and avoiding the baggage of their modern English counterparts.

A

Aisthēsis (Αἴσθησις)

- **Literal Meaning:** Perception, sensation.

- **NPN Definition:** The foundational stage of the epistemic process; the direct, causal interaction between the *Archē* and the nervous system, providing raw sensory data. It is the non-negotiable empirical anchor for all knowledge, though not considered infallible.

---

22. Attributed to Socrates. This aphorism is widely cited in the Socratic tradition, though likely a paraphrase of his student Antisthenes, who said, "The beginning of education is the examination of names" (Epictetus, *Discourses*, 1.17.12).

23. Aristotle, *Nicomachean Ethics*, 1094b23–25, in *The Complete Works of Aristotle: The Revised Oxford Translation*, ed. Jonathan Barnes (Princeton, NJ: Princeton University Press, 1984).

Apeiron (Ἄπειρον)

- **Literal Meaning:** The boundless, unlimited, indefinite.

- **NPN Definition:** The fundamental and permanently unknowable context for the *Archē*. It is the absolute limit of the *Logos*, representing the indeterminate ground from which the determinate *Archē* emerges. Questions that seek to go "outside" or "before" the *Archē* are category errors pointing to the *Apeiron*.

Aporia (Ἀπορία)

- **Literal Meaning:** A state of perplexity, impasse, being at a loss.

- **NPN Definition:** A productive and fertile state of intellectual void induced by the Socratic *Elenchus*. It is the conscious recognition of a flawed model and the lack of a current path forward. It is not a failure but the successful outcome of cognitive *Katharsis* and the essential starting point for genuine inquiry.

Archē (Ἀρχή)

- **Literal Meaning:** Beginning, origin, first principle.

- **NPN Definition:** The sole fundamental existence; the closed, causal system of objective, physical reality. It is a hylomorphic totality, constituted by *Hylē* (dynamic potentiality) as structured by *Logos* (inherent order). It is not a static state but a dynamic, evolutionary process.

Aretē (Ἀρετή)

- **Literal Meaning:** Excellence, virtue.

- **NPN Definition:** Functional excellence; the capacity to fulfill an entity's inherent function within the lawful whole of the *Archē*. For a human, this is the excellent activity of the Psyche in a state of *Dikaiosynē*.

D

Daimonion (Δαιμόνιον)

- **Literal Meaning:** A divine or spiritual entity.

- **NPN Definition:** (In the Socratic context) Understood within NPN as Socrates' highly attuned **faculty of instinctual logic**–a hardwired, pre-rational attunement to the *Logos*. It acts as a negative guide, an intuition of misalignment that prevents logical or ethical error. In NPN terms this is the innate *Genetic Knowledge* sending signals of inherent misalignment with the *Logos*.

Demiourgos (Δημιουργός)

- **Literal Meaning:** Craftsman, artisan.

- **NPN Definition:** A methodological archetype representing one of the two primary modes of engagement for the Navigator. The *Demiourgos* seeks mastery through structural analysis, decomposition, and the perfection of existing forms and rules. In politics and art, it represents the controller, the perfecter, and the system-builder who works with a known blueprint. *It is crucial to note that these modes are fluid modes of engagement, not fixed identities.*

Dikaiosynē (Δικαιοσύνη)

- **Literal Meaning:** Justice, righteousness.

- **NPN Definition:** The functional harmony of a complex system. In the individual *Psyche*, it is the state where the *Logistikon* rightly governs *Orexis* and *Thymos* under the direction of the *Nous*, leading to sustainable flourishing. It is the objective state of a system (individual or social) that is correctly ordered and aligned with the

Logos, thus successfully navigating the constraints of *Dikē*. It is the ethical achievement of the Navigator. Contrast with *Dikē*.

Dikē (Δίκη)

- **Literal Meaning:** Justice, order, judgment.

- **NPN Definition:** The impersonal, causal justice of the *Archē*; the necessary rebalancing and constraints enforced by the framework itself, as in Anaximander's concept where things "make reparation for their injustice according to the assessment of Time." It represents the objective, amoral consequences inherent in the structure of reality. Contrast with *Dikaiosynē*.

E

Elenchus (Ἔλεγχος)

- **Literal Meaning:** Cross-examination, refutation.

- **NPN Definition:** The primary tool of Socratic therapy; a rigorous, logical interrogation of a person's belief structures. Its purpose is to test internal coherence and expose contradiction, inducing *Aporia* by systematically applying the internal *logos* to break down *Doxa*.

Empeiros (Ἔμπειρος)

- **Literal Meaning:** Experienced, practiced.

- **NPN Definition:** The second stage of the epistemic process; accumulated experience and pattern recognition formed through memory and repetition from *Aisthēsis*. It represents the mind's first successful compressions of the *Logos* into efficient, practical, but often non-verbal and context-dependent rules of thumb.

Glossary of Key Terms

Energeia (Ἐνέργεια)

- **Literal Meaning:** Actuality, being-at-work.

- **NPN Definition:** The active, sustained, and generative expression of a *Nous* that has successfully realigned itself with the *Logos*. It is not a static state of possessing knowledge but the continuous activity of deploying a high-confidence model in the world, characterized by effective action and engagement with the *Archē*.

Epistēmē (Ἐπιστήμη)

- **Literal Meaning:** Knowledge, understanding, science.

- **NPN Definition:** Justified, generic knowledge. The culmination of the epistemic process, representing a structured, abstract, and causal model that explains the *why* behind the patterns of *Empeiros*. It is characterized by genericity, causal explanation, and justification through predictive success. Within NPN, *Epistēmē* is understood as a **confidence gradient**—a high-utility, provisional model, not a state of certain, final truth.

Eudaimonia (εὐδαιμονία)

- **Literal Meaning:** Human flourishing, prosperity, blessedness.

- **NPN Definition:** The objective state of a human system functioning at its peak integrity. It is the output of a *Psyche* in *Energeia*, resulting from a life lived in functional alignment with the *Logos*. It is not merely a feeling of happiness, but the state of flourishing that arises from effective diachronic navigation.

Eukairia (Εὐκαιρία)

- **Literal Meaning:** The right or critical moment, opportunity.

- **NPN Definition:** The precondition of cognitive receptivity. A specific configuration of the *Psyche* where the *Thymos* is not defensive and the *Logistikon* is curious and active. It is the "teachable moment" or fertile ground necessary for the *Elenchus* to begin.

Exaiphnes (Ἐξαίφνης)

- **Literal Meaning:** Suddenly, abruptly.

- **NPN Definition:** The sudden, jarring metacognitive awareness of a model's failure that accompanies the *Lysis*. It is the "Aha!" moment of profound disorientation when one realizes, "What I thought I knew, I do not know."

G

Genetor (Γενέτωρ)

- **Literal Meaning:** Begetter, ancestor, father.

- **NPN Definition:** A methodological archetype representing one of the two primary modes of engagement for the Navigator, opposed to the *Demiourgos*. The *Genetor* seeks creation through synthesis, combination, and the generation of novel forms, models, and frameworks. In politics and art, it represents the pioneer, the cartographer of uncharted territory, and the source of new paradigms. *These modes are not a rigid dichotomy but a dynamic polarity between which the skilled Navigator fluidly moves.*

H

Hormē (Ορμή)

- **Literal Meaning:** An impulse, urge, inclination.
- **NPN Definition:** The constitutive drive for life to survive the lawful structure of reality by changing causal flow to aid survival and flourish. It is the engine of life itself and scale-invariant. In humans,

Glossary of Key Terms

it is expressed through the layered faculties of *Orexis*, *Thymos*, *Logistikon*, and the *Nous*.

- **Scientific Definition:** The continuous, internally regulated thermodynamic work required to maintain a far-from-equilibrium organizational state against entropic dispersion. It is the physical expression of persistence: the necessary activity that constitutes being alive, measurable as the energy flow that sustains bounded low-entropy organization.

Hylē ('Ύλη)

- **Literal Meaning:** Wood, material, matter.

- **NPN Definition:** One of the two interdependent principles of the *Archē*. It is the dynamic, undifferentiated physical potentiality; the "stuff" of the cosmos capable of taking on all forms. It is *Archē* considered as raw material, which is always structured by *Logos*.

K

Katharsis (Κάθαρσις)

- **Literal Meaning:** Purification, cleansing, purgation.

- **NPN Definition:** The intended result of the Socratic *Lysis*. It is the conscious purging or cleansing of the *Psyche*, leaving it in a purified, sterile state of *Aporia*, free from the pollution of unexamined *Doxa* and ready to receive genuine *Epistēmē*. In art, it is the psychological result of a controlled, safe *Lysis*.

L

Logistikon (Λογιστικόν)

- **Literal Meaning:** The calculating or reasoning part.

- **NPN Definition:** The executive function of the *Nous*; the calculating faculty. It emerged as a **necessary processor** to resolve the complex, often contradictory data from *Orexis* and *Thymos* into coherent social action. It does not replace the lower layers of the *Psyche* but orchestrates them into a single, non-contradictory stream of behavior directed toward fulfilling the *Hormē* across time. Its function is to model the long-term *Logos* and strategically deploy or suppress *Orexis* and *Thymos* to achieve goals.

Logos (Λόγος)

- **Literal Meaning:** Word, reason, principle, account.

- **NPN Definition:** One of the two interdependent principles of the *Archē*. It is the inherent, constitutive, discoverable, and impersonal order that structures the *Hylē*. It is the complete set of consistent laws and patterns that govern all behavior and manifestation within the *Archē*, making reality predictable and intelligible. It is the target of all epistemic activity.

logos (λόγος)

- **Literal Meaning:** Word, reason, principle, account.

- **NPN Definition:** The internal reasoning faculty or evolved logic of the individual mind (*Nous*). This term is used in lower-case to distinguish the subjective human model (the "map") from the external cosmic law (*Logos*). **See *Somatic logos*.**

Lysis (Λύσις)

- **Literal Meaning:** A releasing, loosening, dissolution.

- **NPN Definition:** The dissolution or breaking down of a rigid, flawed cognitive structure (*Doxa*). It is the destructive, necessary first stage of Socratic therapy and cognitive transformation, where foundational assumptions are dissolved, leading to the state of *Aporia*.

N

Navigator (Ναυτικός) - (Two Tiers)

- **navigator (generic):** The emergent *Nous* in its functional capacity as a model-builder and striver. This is the base state of the human cognitive system, which may be operating on unexamined *Doxa*.

- **Navigator (actualized):** The ideal state of an individual whose *Nous* has achieved meta-cognitive self-awareness and the capacity for conscious *Prohairesis*. The Navigator is not defined by a fixed identity as a *Demiourgos* or *Genetor*, but by the *Phronēsis* (practical wisdom) to consciously discern and walk the path demanded by the specific challenge within the *Archē*. The Navigator is the emergent human capacity for diachronic model-building and alignment, fully realized in a state of *Energeia*.

Neikos (Νεῖκος)

- **Literal Meaning:** Strife, quarrel.

- **NPN Definition:** A fundamental, impersonal cosmic force identified by Empedocles and grounded in Anaximander's cosmic justice. It is the principle of separation, division, repulsion, and individuation. It drives differentiation, competition, and the maintenance of boundaries, manifesting in evolution, entropy, and strategic conflict.

Nous (Νοῦς)

- **Literal Meaning:** Mind, intellect, understanding.

- **NPN Definition:** The faculty of consciousness concerned with insight, intellect, and understanding. Within NPN, it is not a mysterious substance but a natural, **emergent phenomenon**–the Navigator. It is the capacity for abstract modeling, a functional, evolved property of a complex system (the brain) that allows for diachronic navigation of the *Archē*.

O

Orexis (Ὄρεξις)

- **Literal Meaning:** Desire, appetite.

- **NPN Definition:** The most ancient layer of the *Psyche*, the domain of immediate instinct. It is the direct biochemical expression of the *Hormē* for immediate survival and homeostatic regulation, operating on a very short-term timeframe through drives like hunger, thirst, and fear.

P

Philia (Φιλία)

- **Literal Meaning:** Friendship, love, affection.

- **NPN Definition:** A fundamental, impersonal cosmic force identified by Empedocles. It is the principle of attraction, union, integration, and cohesion. It drives combination, cooperation, and the formation of new, more complex wholes, manifesting in gravity, chemical bonding, symbiosis, and social affiliation.

Physis (Φύσις)

- **Literal Meaning:** Nature, the natural order.

- **NPN Definition:** The essential, emergent nature and origin of all things. The object of Pre-Platonic inquiry, which NPN seeks to recover. It is the *Archē* understood as a dynamic, emergent process (*Hylē*-as-structured-by-*Logos*), rather than a static collection of objects.

Prohairesis (Προαίρεσις)

- **Literal Meaning:** Deliberate choice, moral purpose.

- **NPN Definition:** The fundamental methodological choice in inquiry, made from the state of *Dikaiosynē*. It is a deliberate commitment to one of two modes for engaging the unknown: **Mode A (Structural Analysis / *Demiourgos*)** for decomposition and mastery, or **Mode B (Synthesis / *Genetor*)** for the construction of novel models and artifacts. This is the conscious choice, contrasted with *Phronēsis*.

Phronēsis (Φρόνησις)

- **Literal Meaning:** Practical wisdom, prudence.

- **NPN Definition:** The meta-cognitive capacity of the Navigator to discern which mode of engagement (*Demiourgos* or *Genetor*) is demanded by a specific challenge within the *Archē*. It is the operational intelligence that applies general principles (*Logos*) to particular, dynamic situations. Unlike theoretical knowledge (*Epistēmē*), *Phronēsis* is expressed in timely, effective action and is cultivated through experience and reflective practice. This is the unconscious, automated discernment, contrasted with *Prohairesis*.

Psyche (Ψυχή)

- **Literal Meaning:** Soul, life-force, mind.

- **NPN Definition:** The stratified, quadripartite structure of the human mind. It is not a ghost in the machine but an evolutionary layering of the core *Hormē*, consisting of *Orexis* (instinct), *Thymos* (social strategy), and the emergent *Nous* governed by the *Logistikon*. Its proper state is *Dikaiosynē*.

S

Somatic logos (Σωματικός Λόγος)

- **Literal Meaning:** The logic of the body.

- **NPN Definition:** The reasoning capacity of the *Nous*, understood as an evolved heuristic inextricably tied to our physical, evolutionary history. The "laws of logic" are not glimpses of a transcendent realm but highly reliable rules for modeling the macroscopic world, hardwired into our neural architecture because they were evolutionarily successful.

Synthesis (Σύνθεσις)

- **Literal Meaning**: A putting together, composition, combination.

- **NPN Definition:** The core methodological operation of the Genetor mode; synonymous with Coagula. It is the creative act of combining, re-integrating, or "coagulating" existing elements, models, or data to generate a novel, emergent whole. This is the process of building new maps of the Archē where none existed before, producing new paradigms, frameworks, and artifacts. It is the constructive counterpart (Coagula) to the analytical decomposition (Solve) of the Demiourgos.

T

Thymos (Θυμός)

- **Literal Meaning:** Spirit, passion, anger.

- **NPN Definition:** The mid-layer of the *Psyche*, the strategic social calculator. It manages the complex in-group/out-group dynamics of *Philia* and *Neikos*, translating these cosmic forces into social emotions like anger, courage, pride, shame, and loyalty, all in service of the individual's and group's *Hormē*.

Notes on the Method: AI and the Future

"The map is not the territory." — Alfred Korzybski

The Technological Scaffold

The construction of a systematic philosophy outside the traditional structures of the academy presents a profound challenge: the absence of a community of unbiased, capable evaluators to stress-test a novel, fully interlocked structure. To bridge this gap, Large Language Models were integrated into the drafting process.[24]

I employed these tools not as sources of ideas, but as instruments for a "logic and rigor check"—a mechanism to convert mental construction into linguistic formality. They acted as a rigorous *Elenchus* to my *Doxa*, forcing linguistic precision where there was previously only conceptual intuition. They provided the necessary mirroring that allowed the structure to be built in words.

Furthermore, they served as a cognitive "force multiplier," overcoming inherent human limitations. The laborious tasks of turning thought into text—grammar, spelling, editing, and the sheer physical act of typing—were drastically accelerated. This allowed me to remain in a sustained state of focused ideation, acting as the "idea engine" while the AI served as a "translation and articulation engine." It is not an accident that a work of this scope emerged at the dawn of the LLM era; the tool provided the necessary leverage to exteriorize a system of this complexity.

---

24. AI language models were employed as exploratory and editorial aids in the development of this system, with due regard for their epistemological limitations. Cf. Emily M. Bender et al., "On the Dangers of Stochastic Parrots: Can Language Models Be Too Big?," FAccT '21 (2021): 610–623.

The AI also excelled at generating coherent metaphors and analogies, many of which were adopted to aid reader comprehension. It assisted immensely in maintaining a consistent professional tone and structural layout—an invaluable aid for an outsider to academic publishing.

Yet, for all its utility, the AI was a tool, not a partner. It consistently exhibited conceptual drift, requiring me to hold the definitions of all core concepts with absolute accuracy in my own mind to constantly correct its missteps. The primary connections, the foundational insights, and the entire architectural blueprint are mine. The LLM did not generate the thesis; it helped me build the edifice from the blueprint I had spent a decade drawing in the *Nous* of a Navigator.

Acknowledgments

"We are like dwarfs on the shoulders of giants, so that we can see more than they..."
— Bernard of Chartres

No philosophy is born in a vacuum. This system is a reconstruction, built upon the foundations laid by twenty-five centuries of thinkers who worked tirelessly to advance the frontier of human understanding.

To the architects of the past—from the first *physiologoi* who dared to measure the earth, to the logicians who mapped the mind, and the scientists who charted the *Archē*—we owe the very ground we stand on. Their labor was the first navigation; this work is merely a restoration of their maps.

Finally, gratitude is due to you, the reader. A text is inert until it is engaged by a living *Nous*. By dedicating your time and rigorous attention to these ideas, you cease to be a spectator of this tradition and become a participant in its unfolding.

The project of human thought is unfinished. It continues with you.

References

Agamben, Giorgio. *The Open: Man and Animal.* Translated by Kevin Attell. Stanford: Stanford University Press, 2004.

Alexandria, Pappus of. *Synagoge.* Book VIII. Cited historically.

Annas, Julia. *An Introduction to Plato's Republic.* Oxford: Clarendon Press, 1981.

———. *Platonic Ethics, Old and New.* Ithaca, NY: Cornell University Press, 1999.

———. *Virtue and Law in Plato and Beyond.* Oxford: Oxford University Press, 2017.

Attwell, David, and Simon B. Laughlin. "An Energy Budget for Signaling in the Grey Matter of the Brain." *Journal of Cerebral Blood Flow & Metabolism* 21, no. 10 (2001): 1133–1145.

Baars, Bernard J. *A Cognitive Theory of Consciousness.* Cambridge: Cambridge University Press, 1988.

Barnes, Jonathan, ed. *The Complete Works of Aristotle: The Revised Oxford Translation.* Includes: Metaphysics, De Anima, Physics, Nicomachean Ethics, and Politics. Princeton, NJ: Princeton University Press, 1984.

———. *The Presocratic Philosophers.* London: Routledge / Kegan Paul, 1982.

Barsalou, Lawrence W. "Perceptual Symbol Systems." *Behavioral and Brain Sciences* 22, no. 4 (1999): 577–660.

Bender, Emily M., Timnit Gebru, Angelina McMillan-Major, and Shmargaret Shmitchell. "On the Dangers of Stochastic Parrots: Can Language Models Be Too Big?" In *Proceedings of the 2021 ACM Conference on Fairness, Accountability, and Transparency*, 610–623. New York, NY: ACM, 2021.

Berg, Howard C. *Random Walks in Biology*. Expanded. Princeton: Princeton University Press, 1993.

Bergson, Henri. *Creative Evolution*. Translated by Arthur Mitchell. New York: Henry Holt / Company, 1911.

———. *Duration and Simultaneity: Bergson and the Einsteinian Universe*. Translated by Leonora Alvès. Manchester, UK: Clinamen Press, 1998.

Berlin, Isaiah. "Two Concepts of Liberty." In *Four Essays on Liberty*, 118–172. Oxford: Oxford University Press, 1969.

Betegh, Gábor. *The Derveni Papyrus: Cosmology, Theology and Interpretation*. Cambridge: Cambridge University Press, 2004.

Bodhi, Bhikkhu, trans. *The Connected Discourses of the Buddha: A Translation of the Saṃyutta Nikāya*. Boston: Wisdom Publications, 2005.

———, trans. *The Numerical Discourses of the Buddha: A Translation of the Aṅguttara Nikāya*. Boston: Wisdom Publications, 2012.

Brentano, Franz. *The Psychology of Aristotle: In Particular His Doctrine of the Active Intellect*. Translated by Rolf George. Berkeley: University of California Press, 1977.

Buddharakkhita, Acharya, trans. *The Dhammapada: The Buddha's Path of Wisdom*. Kandy: Buddhist Publication Society, 1985.

Burnyeat, Myles. *The Theaetetus of Plato*. Indianapolis, IN: Hackett Publishing, 1990.

Chalmers, David J. *The Conscious Mind: In Search of a Fundamental Theory*. New York: Oxford University Press, 1996.

Churchland, Patricia S. *Touching a Nerve: The Self as Brain.* New York: W. W. Norton & Company, 2013.

Confucius. *Analects.* Cited by book and verse, e.g., 2.3.

Cooper, John M., ed. *Plato: Complete Works.* Includes: Apology, Euthyphro, Meno, Phaedo, Republic, Symposium, Theaetetus, Parmenides, Sophist, Philebus, Timaeus, and Laws. Indianapolis, IN: Hackett Publishing Company, 1997.

Copleston, Frederick. *A History of Philosophy.* 9 vols. London: Burns, Oates & Washbourne, 1946–1975.

Cornford, Francis MacDonald. *Plato's Theory of Knowledge: The Theaetetus and the Sophist.* London: Routledge & Kegan Paul, 1935.

Cracraft, Joel. "The species of the birds-of-paradise (Paradisaeidae): applying the phylogenetic species concept to a complex pattern of diversification." *Cladistics* 8, no. 1 (1992).

Curd, Patricia. "Presocratic Philosophy." In *The Stanford Encyclopedia of Philosophy*, Summer 2012, edited by Edward N. Zalta. Metaphysics Research Lab, Stanford University, 2012.

Damasio, Antonio. *Descartes' Error: Emotion, Reason, and the Human Brain.* New York: G. P. Putnam's Sons, 1994.

Darwin, Charles. *On the Origin of Species by Means of Natural Selection.* London: John Murray, 1859.

Dawkins, Richard. *The Selfish Gene.* Oxford: Oxford University Press, 1976.

———. *The Selfish Gene.* 30th anniversary. Oxford: Oxford University Press, 2006.

Deacon, Terrence W. *Incomplete Nature: How Mind Emerged from Matter.* New York: W. W. Norton & Company, 2012.

Dehaene, Stanislas. *Consciousness and the Brain: Deciphering How the Brain Codes Our Thoughts.* New York: Viking, 2014.

Dehaene, Stanislas. *How We Learn: Why Brains Learn Better Than Any Machine...For Now*. New York: Viking, 2020.

Dennett, Daniel C. *Darwin's Dangerous Idea: Evolution and the Meanings of Life*. New York: Simon & Schuster, 2014.

———. *Freedom Evolves*. New York: Viking, 2003.

———. *From Bacteria to Bach and Back: The Evolution of Minds*. New York: W. W. Norton & Company, 2017.

———. *Kinds of Minds: Toward an Understanding of Consciousness*. New York: Basic Books, 1996.

Descartes, René. *Meditations on First Philosophy*, translated by John Cottingham, Robert Stoothoff, and Dugald Murdoch. Cambridge: Cambridge University Press, 1984.

Deutscher, Eli Adam. "Anaximander and the Zero Principle: The Relational Ontology of the Apeiron." *Preprint*, 2026. https : / / neopreplatonic.com/papers/anaximander.

———. "Parmenides the Polemicist: The Eleatic Crisis and the Indeterminate Ground of Thought." Preprint, 2026. https : / / www . neopreplatonic.com/papers/Parmenides/.

Dewey, John. *Experience and Nature*. Chicago: Open Court, 1925.

———. *The Quest for Certainty: A Study of the Relation of Knowledge and Action*. New York: Minton, Balch & Company, 1929.

Diels, Hermann, and Walther Kranz, eds. *Die Fragmente der Vorsokratiker*. 6th. Vol. 1. Berlin: Weidmann, 1951.

Dissanayake, Ellen. *Homo Aestheticus: Where Art Comes From and Why*. New York: The Free Press, 1992.

Duncker, Karl. "On Problem-Solving." *Psychological Monographs* 58, no. 5 (1945).

Durkheim, Emile. *The Division of Labor in Society*. Translated by George Simpson. New York: The Free Press, 1997.

Dutton, Denis. *The Art Instinct: Beauty, Pleasure, and Human Evolution*. New York: Bloomsbury Press, 2009.

England, Jeremy L. "Dissipative Adaptation in Driven Self-Assembly." *Nature Nanotechnology* 10, no. 11 (2015): 919–923.

Fink, Bernhard, and Nick Neave. "The biology of facial beauty." *International Journal of Cosmetic Science* 27, no. 6 (2005): 317–325.

Flanagan, Owen. *The Bodhisattva's Brain: Buddhism Naturalized*. Cambridge, MA: MIT Press, 2011.

Franzén, Torkel. *Gödel's Theorem: An Incomplete Guide to Its Use and Abuse*. Wellesley, MA: A K Peters, 2005.

Friston, Karl. "The Free-Energy Principle: A Unified Brain Theory?" *Nature Reviews Neuroscience* 11, no. 2 (2010): 127–138.

Furley, David J. *The Greek Cosmologists: Volume 1, The Formation of the Atomic Theory and its Earliest Critics*. Cambridge: Cambridge University Press, 1987.

Gardner, Daniel K. *Learning to Be a Sage: Selections from the Conversations of Master Zhu, Arranged Topically*. Trans. of Zhuzi yulei by Zhu Xi. Berkeley: University of California Press, 1990.

Gödel, Kurt. "Über formal unentscheidbare Sätze der *Principia Mathematica* und verwandter Systeme I." English translation: "On Formally Undecidable Propositions of *Principia Mathematica* and Related Systems I", *Monatshefte für Mathematik und Physik* 38, no. 1 (1931): 173–198.

Gould, Stephen Jay, and Richard C. Lewontin. "The Spandrels of San Marco and the Panglossian Paradigm: A Critique of the Adaptationist Programme." *Proceedings of the Royal Society of London B* 205, no. 1161 (1979): 581–598.

Grammer, Karl, and Randy Thornhill. "Human (Homo sapiens) facial attractiveness and sexual selection: The role of symmetry and averageness." *Journal of Comparative Psychology* 108, no. 3 (1994): 233–242.

Gregory, Andrew. *Anaximander: A Re-assessment*. London: Bloomsbury Academic, 2016.

Griffin, James. *Well-Being: Its Meaning, Measurement, and Moral Importance*. Oxford: Clarendon Press, 1986.

Guthrie, W. K. C. *A History of Greek Philosophy, Vol. 1*. Cambridge: Cambridge University Press, 1962–1969.

———. *A History of Greek Philosophy, Vol. 4*. Cambridge: Cambridge University Press, 1975.

———. *A History of Greek Philosophy, Vol. 5*. Cambridge: Cambridge University Press, 1978.

———. *A History of Greek Philosophy, Vol. 6: Aristotle: An Encounter*. Cambridge: Cambridge University Press, 1981.

Hadot, Pierre. *Philosophy as a Way of Life: Spiritual Exercises from Socrates to Foucault*. Translated by Michael Chase. Oxford: Blackwell, 1995.

Haidt, Jonathan. *The Righteous Mind: Why Good People Are Divided by Politics and Religion*. New York: Pantheon Books, 2012.

Hamilton, W. D. "The Genetical Evolution of Social Behaviour, I and II." *Journal of Theoretical Biology* 7, no. 1 (1964): 1–52.

Hayek, F. A. *Law, Legislation and Liberty, Vol. 1: Rules and Order*. Chicago: University of Chicago Press, 1973.

Hayes, J. R. *The Complete Problem Solver*. 2nd. Hillsdale, NJ: Erlbaum, 1989.

Heidegger, Martin. *Being and Time*. Translated by John Macquarrie and Edward Robinson. New York: Harper & Row, 1962.

Heisenberg, Werner. *Physics and Philosophy: The Revolution in Modern Science.* New York: Harper & Row, 1958.

———. "The Physical Content of Quantum Kinematics and Mechanics." Original German title: Über den anschaulichen Inhalt der quantentheoretischen Kinematik und Mechanik. English trans. in Quantum Theory and Measurement, *Zeitschrift für Physik* 43 (1927): 172–198.

Hempel, Carl G. *Philosophy of Natural Science.* Englewood Cliffs, NJ: Prentice-Hall, 1966.

Hilbert, David. *Grundlagen der Geometrie.* English translation: *Foundations of Geometry.* Leipzig: B. G. Teubner, 1899.

Hughes, Susan M., and Gordon G. Gallup. "Sex differences in morphological predictors of sexual behavior: Shoulder to hip and waist to hip ratios." *Evolution and Human Behavior* 24, no. 3 (2003): 173–178.

Hume, David. *A Treatise of Human Nature.* Edited by David Fate Norton and Mary J. Norton. Oxford: Oxford University Press, 2007.

Husserl, Edmund. *The Crisis of European Sciences and Transcendental Phenomenology: An Introduction to Phenomenological Philosophy.* Evanston, IL: Northwestern University Press, 1970.

Irwin, Terence. *Plato's Ethics.* Oxford: Oxford University Press, 1995.

James, William. *Pragmatism: A New Name for Some Old Ways of Thinking.* New York: Longmans, Green, / Co., 1907.

Juarrero, Alicia. *Dynamics in Action: Intentional Behavior as a Complex System.* Cambridge, MA: MIT Press, 1999.

Kahn, Charles H. *Anaximander and the Origins of Greek Cosmology.* New York: Columbia University Press, 1960.

———. *The Art and Thought of Heraclitus: An Edition of the Fragments with Translation and Commentary.* Cambridge: Cambridge University Press, 1979.

Kahneman, Daniel. *Thinking, Fast and Slow.* New York: Farrar, Straus / Giroux, 2011.

Kant, Immanuel. *Critique of Practical Reason.* Edited and translated by Mary Gregor. Cambridge Texts in the History of Philosophy. Cambridge: Cambridge University Press, 1997.

———. *Critique of Pure Reason.* Translated by Paul Guyer and Allen W. Wood. Originally published 1781/1787. Cambridge: Cambridge University Press, 1998.

Kirk, G. S., J. E. Raven, and M. Schofield. *The Presocratic Philosophers: A Critical History with a Selection of Texts.* 2nd. Cambridge: Cambridge University Press, 1983.

Kirton, M. J. "Adaptors and Innovators: A Description and Measure." *Journal of Applied Psychology* 61, no. 5 (1976): 622–629.

Kraut, Richard, ed. *The Cambridge Companion to Plato.* Cambridge: Cambridge University Press, 1992.

Kuhn, Thomas S. *The Structure of Scientific Revolutions.* 2nd. Chicago: University of Chicago Press, 1970.

Lakatos, Imre. "Falsification and the Methodology of Scientific Research Programmes." In *Criticism and the Growth of Knowledge*, edited by Imre Lakatos and Alan Musgrave, 91–196. Cambridge: Cambridge University Press, 1970.

Lakoff, George, and Mark Johnson. *Philosophy in the Flesh: The Embodied Mind and Its Challenge to Western Thought.* New York: Basic Books, 1999.

Laozi. *Tao Te Ching.* Cited by chapter, e.g., Ch. 60.

Lawrence, Gavin. "The Aporia of the *Meno* and the Eristic of the *Euthydemus.*" In *Maiensis: Essays in Ancient Philosophy in Honour of Myles Burnyeat*, edited by Dominic Scott. Oxford: Oxford University Press, 2007.

Lear, Jonathan. "Katharsis." *Phronesis* 33, no. 3 (1988): 297–326.

Legge, James, trans. *The Chinese Classics. Vol. 3: The Shoo King (Shujing)*. London: Trübner & Co., 1865.

Lesher, J. H. *Xenophanes of Colophon: Fragments*. Toronto: University of Toronto Press, 1992.

Lloyd, G. E. R. *Methods and Problems in Greek Science*. Cambridge: Cambridge University Press, 1991.

———. *Polarity and Analogy: Two Types of Argumentation in Early Greek Thought*. Cambridge: Cambridge University Press, 1966.

Lloyd, G. E. R., and Nathan Sivin. *The Way and the Word: Science and Medicine in Early China and Greece*. New Haven: Yale University Press, 2002.

MacIntyre, Alasdair. *After Virtue*. 3rd. Notre Dame: University of Notre Dame Press, 2007.

Maturana, Humberto R., and Francisco J. Varela. *Autopoiesis and Cognition: The Realization of the Living*. Dordrecht: D. Reidel Publishing Co., 1980.

Mayr, Ernst. *Systematics and the Origin of Species*. New York: Columbia University Press, 1942.

McKirahan, Richard D. *Philosophy Before Socrates: An Introduction with Texts and Commentary*. 2nd. Indianapolis, IN: Hackett Publishing Company, 2010.

Mencius. *Mencius*. Cited by book and verse, e.g., 1B.8.

Merleau-Ponty, Maurice. *Phenomenology of Perception*. Translated by Colin Smith. London: Routledge & Kegan Paul, 1962.

Mill, John Stuart. *On Liberty*. London: John W. Parker / Son, 1859.

Millikan, Ruth Garrett. *Language, Thought, and Other Biological Categories.* Cambridge, MA: MIT Press, 1984.

Nagel, Ernest, and James R. Newman. *Gödel's Proof.* Revised. New York: New York University Press, 2001.

Ñāṇamoli, Bhikkhu, and Bhikkhu Bodhi, trans. *The Middle Length Discourses of the Buddha: A Translation of the Majjhima Nikāya.* Boston: Wisdom Publications, 1995.

Nietzsche, Friedrich. *Beyond Good and Evil.* Translated by Walter Kaufmann. New York: Vintage Books, 1966.

———. *The Will to Power.* Edited by Walter Kaufmann. New York: Vintage, 1968.

Nozick, Robert. *Anarchy, State, and Utopia.* New York: Basic Books, 1974.

Nussbaum, Martha C. *Creating Capabilities: The Human Development Approach.* Cambridge, MA: Harvard University Press, 2011.

———. *The Fragility of Goodness: Luck and Ethics in Greek Tragedy and Philosophy.* Cambridge: Cambridge University Press, 1986.

———. *The Therapy of Desire: Theory and Practice in Hellenistic Ethics.* Princeton, NJ: Princeton University Press, 1994.

Nyquist, Mary. *Arbitrary Rule: Slavery, Antislavery, and the Roots of American Freedom.* Chicago: University of Chicago Press, 2013.

Pinker, Steven. *How the Mind Works.* New York: W. W. Norton & Company, 1997.

———. *The Blank Slate: The Modern Denial of Human Nature.* New York: Viking, 2002.

Popper, Karl. *The Logic of Scientific Discovery.* London: Routledge, 1982.

———. *The Open Society and Its Enemies, Vol. 1.* Princeton, NJ: Princeton University Press, 1945.

Putnam, Hilary. *Reason, Truth and History.* Cambridge: Cambridge University Press, 1981.

Raichle, Marcus E. "The Brain's Dark Energy." *Science* 314, no. 5803 (2006): 1249–1250.

Railton, Peter. "Moral Realism." *The Philosophical Review* 95, no. 2 (1986): 163–207.

Rawls, John. *A Theory of Justice.* Cambridge, MA: Belknap Press of Harvard University Press, 1971.

Reeve, C.D.C. *Philosopher-Kings: The Argument of Plato's Republic.* Princeton, NJ: Princeton University Press, 1988.

Rescher, Nicholas. *Process Metaphysics: An Introduction to Process Philosophy.* Albany, NY: State University of New York Press, 1996.

Reynolds, David S. *Mightier Than the Sword: Uncle Tom's Cabin and the Battle for America.* New York: W. W. Norton & Company, 2011.

Roochnik, David. "What Is Natural about Aristotle's Polis?" *Polis* 26, no. 2 (2009): 217–240.

Sayre, Kenneth M. *Plato's Late Ontology: A Riddle Resolved.* Princeton: Princeton University Press, 1983.

Schofield, Malcolm. "Ideology and Philosophy in Aristotle's Theory of Slavery." In *Aristoteles' "Politik": Akten des XI. Symposium Aristotelicum,* edited by Günther Patzig, 1–27. Göttingen: Vandenhoeck & Ruprecht, 1990.

Schrödinger, Erwin. *What is Life? The Physical Aspect of the Living Cell.* Cambridge: Cambridge University Press, 1944.

Scott, Dominic. *Recollection and Experience: Plato's Theory of Learning and Its Successors.* Cambridge: Cambridge University Press, 1995.

Searle, John R. "Minds, Brains, and Programs." *Behavioral and Brain Sciences* 3, no. 3 (1980): 417–457.

Shenhav, Amitai, Matthew M. Botvinick, and Jonathan D. Cohen. "The Expected Value of Control: An Integrative Theory of Anterior Cingulate Cortex Function." *Neuron* 79, no. 2 (2013): 217–240.

Singh, Devendra. "Adaptive significance of female physical attractiveness: Role of waist-to-hip ratio." *Journal of Personality and Social Psychology* 65, no. 2 (1993): 293–307.

Solmsen, Friedrich. "Aristotle's Word for Matter." *Didascaliae*, 1961, 395–408.

Spencer-Brown, George. *Laws of Form*. London: George Allen / Unwin, 1969.

Spinoza, Benedict de. *Ethics*, translated by Edwin Curley. Princeton, NJ: Princeton University Press, 1994.

———. *Spinoza: Complete Works*. Edited by Michael L. Morgan. Translated by Samuel Shirley. Standard modern reference for Letter 50. Indianapolis, IN: Hackett Publishing Company, 2002.

Tononi, Giulio. "Consciousness as Integrated Information: A Reconstruction." *Biological Bulletin* 215, no. 3 (2008): 216–242.

Trivers, Robert L. "The Evolution of Reciprocal Altruism." *The Quarterly Review of Biology* 46, no. 1 (1971): 35–57.

Vaihinger, Hans. *The Philosophy of 'As If': A System of the Theoretical, Practical and Religious Fictions of Mankind*. Translated by C. K. Ogden. London: Kegan Paul, Trench, Trubner & Co., 1924.

Vernant, Jean-Pierre. *The Origins of Greek Thought*. Ithaca, NY: Cornell University Press, 1982.

Vlastos, Gregory. *Socrates: Ironist and Moral Philosopher*. Ithaca, NY: Cornell University Press, 1991.

———. "The Socratic Elenchus." *Oxford Studies in Ancient Philosophy* 1 (1983): 27–58.

———. "The Third Man Argument in the *Parmenides*." *The Philosophical Review* 63, no. 3 (1954): 319–349.

Wheeler, J. A., and W. H. Zurek, eds. *Quantum Theory and Measurement*. Princeton, NJ: Princeton University Press, 1983.

Whitehead, Alfred North. *Process and Reality*. New York: The Free Press, 1978.

Wilhelm, Richard, trans. *The I Ching or Book of Changes*. Princeton, NJ: Princeton University Press, 1950.

Index

A

Adikia, *see* Logic of Distinction, Adikia (The Cut)
Agency
    Active Boundary, 289
    Conatus (Spinoza), 49
    Concept, 51–53, 57, 60, 61, 63, 64, 76, 83, 89, 90, 166, 193, 194, 199, 251, 267, 268, 290
    Constitutive Striving, 4
    Emergent Agent, 18, 49, 52–58, 61, 83, 144, 151, 192, 193, 199, 226, 244, 245, 268, 284, 287–292, 294
    Free Will vs Determinism, 10, 15, 272
    Hierarchy, 101
    Navigation
        Self-correction, 271
    Passive Boundary, 289
    Pathology
        Internal Conflict, 162, 258, 262
        Internal Misalignment, 258, 259, 262
        Rebellious Agent, 259
    Striving (Hormē), 20, 49, 51–54, 56–58, 61–63, 66, 76, 89, 90, 112, 192, 194, 199, 226, 267, 268, 287, 291, 294
    Teleological Bias, 55
    Teleology, 55, 64, 144, 194
    Will to Power (Nietzsche), 64, 269
Aisthēsis, 62, 98, 101
Allegory of the Cave, 113, 114, 117
Anamnesis, 29, 109, 110, 116
Anattā (Non-self), 260, 262
Anaximander, 9–13, 17, 18, 20, 25, 28, 35, 37, 43, 97, 108, 112, 128, 133, 178, 181, 182, 184, 185, 203–209, 212, 216, 217, 219, 222, 248, 280, 294, 300, 305
Annas, Julia, 28, 114, 127, 130, 150, 162
Apeiron
    epistemology, 12, 85, 88
    First Principle, *see* First Principles, FP5:

Impotence Before
      Apeiron
   metaphysics, 33
   navigation, 147
   Presocratics, 9, 12
Aporia
   epistemology, 21
   navigation, 94, 96, 146
   Plato, 107
   Socrates, 26, 27
Arche
   First Principle, *see* First
      Principles, FP1:
      Primacy of the Arche
Archē
   epistemology, 9, 11, 79, 80
   Hylomorphism
      Computer Analogue, 38
   lexicon, 5, 297
   metaphysics, 31, 34, 36
   politics, 149, 150
Aristotle, 11–13, 52, 99, 102, 104,
      125, 126, 131, 135–145,
      147, 149, 150, 165–167,
      173, 175, 178, 179, 185,
      187, 203, 207, 208,
      217–223, 253, 278, 293,
      296, 297
   Biology
      Scala Naturae, 219
   Critique
      Synchronic Flattening,
         135, 143

Dynamis (Potentiality), 136,
   217
Entelecheia, 137, 138
Faculties of Soul
   Intellectual
      (Dianoētikon), 137
   Nutritive (Threptikon),
      137
   Perceptive (Aisthētikon),
      137
Four Causes, 138, 139, 142
Hylomorphism, 141, 296
Metaphysics
   Energeia vs. Dynamis,
      135–137
   Unmoved Mover, 135,
      141, 175, 178, 179, 220
   Nous Pathētikos (Passive
      Intellect), 140
   Nous Poietikos (Active
      Intellect), 140–142
Ousia (Substance), 136, 145,
   207, 217, 278
Political Theory
   Natural Slave Doctrine,
      144
   Polity (Mixed
      Constitution), 144
Telos
   Eudaimonia, 138
Works
   De Anima, 137, 140
   Nicomachean Ethics, 99,
      102, 142, 296, 297

INDEX

Autopoiesis, 52

B
Barnes, Jonathan, 28, 52, 135, 149, 165, 175, 187, 205, 207, 217, 278, 297
Berg, Howard C., 53
Bergson, Henri, 49, 82
Boundary Condition, see Logic of Distinction
Boundary Maintenance, see First Principles, FP6: Primacy of Hormē
Buddha, 257–259, 262, 263
Buddhism, 257, 258, 263

C
Chalmers, David, 270
Chemotaxis, 199
Confidence Gradient
    Corollary, see Corollaries, C1: Confidence Gradient
    definition, 1, 71, 123
    epistemology, 79, 80, 296
Confucius, 254
Consciousness
    Hard Problem, 65, 270, 271
Corollaries
    C1: Apeironic Context, 36, 250
    C2: Confidence Gradient, 2, 71, 80, 86–89, 182, 192, 195, 196, 206, 215, 225, 226, 228–232, 234–236, 239, 241, 242, 266, 267, 272–274, 288, 301
    C3: Somatic Present, 72, 80, 83, 84, 197
    C4: Objectivity of Value, 54, 57, 60, 62, 76, 197
Cost of Being, see Theorems, T7: The Entropic Asymmetry

D
Daimonion, 26
Daoism, 179, 254
Darwin, Charles, 46, 48, 109, 187, 227
Deacon, Terrence, 55, 64, 221
Dehaene, Stanislas, 72, 111
Demiourgos (Mode A)
    Artistic (Craftsman), 168, 169
    Methodological (Analysis), 99, 145
    Political (State Craft), 150, 152, 153
Dennett, Daniel C., 46, 72, 80, 109, 272
Dewey, John, 2, 81, 104, 266
Diachronic Primacy, see First Principles, FP2: Diachronic Primacy
Diachronic Turn, 1, 6, 17, 31, 48, 57, 83, 150, 217, 222, 256, 275
Dikē, 12, 13, 25, 43, 56, 58, 94, 97–99, 115, 166, 204,

205, 207, 208, 240, 245, 255, 294, 300
Dissipative Structure, 54
Doxa
    epistemology, 16
    political, 150
    socratic therapy, 25, 27
Dukkha (Suffering), 258, 259

E

Eightfold Path, 259
Elenchus
    aesthetics, 169
    navigation, 94, 95
    politics, 157, 159
    Socrates, 26
Empedocles, 10, 14, 15, 20, 25, 31, 43, 44, 112, 133, 247, 252, 295, 305, 306
Empeiros, 85, 86, 101, 301
Energeia
    aesthetics, 171
    navigation, 93, 94, 145, 146
    politics, 151, 152
Entropic Asymmetry, *see* Theorems, T7: The Entropic Asymmetry
Entropy, 15, 44, 45, 47, 52–54, 61, 87, 173, 199, 207, 237, 252, 303, 305
    Dispersion, 55, 303
Epistēmē, 27, 39, 86, 121, 122, 139, 173, 192, 250, 260, 301, 303, 307
    Epistemic Loop, 86
    epistemology, 16, 80, 86
    navigation, 97
    Plato, 116
Eros (Drive), 112, 113, 116
Ethics
    Is-Ought Problem, 61
Eudaimonia
    aesthetics, 173
    ethics, 94, 102, 143, 146
    politics, 152, 161
Eukairia (Receptive Moment), 94, 95, 238
Exaiphnes (The Rupture), 95, 96

F

Falsification, 195, 196, 229–231, 242, 272, 273
Feedback Loop, 101, 110, 255
First Principles
    GZP: General Zero Principle, 34, 35, 45, 108, 109, 137, 141, 175–185, 188, 200, 211, 213, 215–223, 238, 250
    ZP: Zero Principle, 12, 13, 18, 25, 35–37, 42–45, 75, 108, 119, 176, 180, 182, 183, 188, 190, 205, 206, 208, 212–214, 216, 220, 248, 249, 256, 280
    FP1: Primacy of the Archē, 39, 69, 71, 119, 122, 133, 184, 191, 239–241
    FP2: Diachronic Primacy, 32, 54, 59, 83, 84, 89,

INDEX

184, 192, 193, 197, 198, 200, 206, 232, 240, 269–271, 274
FP3: Logos and Exhaustive Polarity, 13, 45, 59, 183, 198, 238, 240, 250, 251
FP4: Potential for Nous, 69, 196, 235, 246, 274
FP5: Impotence Before Apeiron, 41, 129, 179, 184, 220, 239, 241, 250, 273
FP6: Primacy of Horme, 53, 57, 59, 198–200, 241, 270, 273, 289
FP7: Somatic logos, 60, 81, 93, 198
FP8: Navigability of Archē, 40, 184
FP9: Meta-Cognitive Potential, 70, 237, 239–242, 271, 290
ZP: Zero Principle (System Identity), 12, 35, 36, 42, 45, 108–110, 120, 177, 183, 188–190, 196, 203, 208, 209, 211, 212, 214–217, 235, 249, 250, 280
Formal Truths, *see* Theorems, T2: Status of Formal Truths
Friction (Social/Cognitive), 14, 59, 60, 139, 161, 198, 217, 237, 241, 242, 245, 254, 258
Frist philosophy: Geometric Proof, 33, 188, 277, 278, 291

G

Genetic Memory, 55, 109, 258, 299
Genetor (Mode B)
    Artistic (Pioneer), 168
    Methodological (Inquiry), 99, 100, 145
    Political (Network State), 152, 154
Guardian (Platonic Class), 114
Guthrie, W.K.C., 15, 24, 103, 117, 126, 131, 133, 135, 218
Gödel, Kurt, 180, 181

H

Hadot, Pierre, 21, 27, 94, 265
Haidt, Jonathan, 95, 259
Hamartia (Tragic Error), 166
Heidegger, Martin, 6
Heisenberg, Werner, 232
Heraclitus, 9, 10, 13, 14, 17, 20, 23, 25, 28, 51, 58, 133, 247, 294
Hormē
    aesthetics, 171
    epistemology, 79, 89
    First Principle, *see* First Principles, FP6: Primacy of Hormē

lexicon, 5, 304
Hume, David, 10, 57, 267, 268

I
Incompleteness Theorems (Gödel), 180, 181
Irwin, Terence, 118
Is-Ought Problem, 52, 56, 58, 267, 291
    Hume's Guillotine, 267, 268

K
Kahneman, Daniel, 67, 70–72, 225
Kant, Immanuel, 7, 71, 268, 269
Katharsis
    aesthetics, 167
    navigation, 96
    Socrates, 24, 27
Kuhn, Thomas, 96, 168

L
Ladder of Love, 112
Laozi, 179, 237, 249, 254, 279
Laws of Form (Spencer-Brown), 181, 182, 277
    Distinction, 182, 277
Logic of Distinction
    Adikia (The Cut), 277
    Boundary Condition, 33, 188, 277, 278
    Boundary Maintenance, 289
    Interstitial Ground, 281–283
    Region R, 279
Logistikon
    politics, 162
    tripartite soul, 82, 89
Logos
    epistemology, 9, 79
    First Principle, *see* First Principles, FP3: Logos and Exhaustive Polarity
    lexicon, 5, 298
    metaphysics, 38
    politics, 149
Lysis
    aesthetics, 167
    epistemology, 21
    navigation, 94, 95, 146
    Socrates, 24

M
Magnesia (City of Laws), 129
Maturana, Humberto R., 52
Mencius, 255
Meta-Cognitive Potential, *see* First Principles, FP9: Meta-Cognitive Potential
Metaphysics
    Circular Reasoning, 35, 175, 176, 184
    Infinite Regress, 33–36, 120, 128, 175, 176, 178, 181, 183–185, 188, 213, 215, 238, 280, 281, 287
Mill, John Stuart, 146, 149, 158
Millikan, Ruth Garrett, 54
Mimesis

aesthetics, 166

N

Natural Slave, 144
Navigability
   First Principle, *see* First Principles, FP8: Navigability of Arche
Navigation
   Feedback Loop (Popper Protocol), 242
Navigator
   epistemology, 15, 19, 20, 79–81
   metaphysics, 32
   navigation, 94, 95, 143, 145
   politics, 149
Navigator Polis, 146, 158, 159, 161, 162, 173
Neikos (Force of Separation)
   Artistic, 170
   Cognitive, 14
   Cosmic, 31, 43
   Social, 149, 150
   Social Friction, 152, 158
Nietzsche, Friedrich, 64, 226, 269, 293
Noble Lie, 114, 156
Nous
   epistemology, 15, 16, 80, 81
   First Principle, *see* First Principles, FP4: Potential for Nous
   lexicon, 5, 299
   metaphysics, 32, 42
   navigation, 93, 95, 143, 145
Nussbaum, Martha, 20, 24, 29, 97, 102, 127, 138, 155

O

Objectivity of Value Corollary, *see* Corollaries, C3: Objectivity of Value
Orexis
   aesthetics, 167
   navigation, 102
   tripartite soul, 89

P

Parmenides, 10, 15, 16, 20, 32, 80, 85, 118–121, 124, 128, 131, 132, 178, 208, 213, 215, 253, 283, 295
Philia (Force of Union)
   Artistic, 169, 170
   Cognitive, 14
   Cosmic, 31, 43
   Social, 149, 150
Philosopher-King, 113–115, 117, 129, 159
Physis, 10, 13–15, 18–20, 24, 28, 41, 47, 76, 82–84, 86, 87, 101, 103, 111, 118, 123, 124, 130, 135, 136, 138, 147, 161, 166, 174, 219, 226, 270
   lexicon, 7
   metaphysics, 32, 33
   polis, 152, 159

Pinker, Steven, 63, 109, 258
Plato, 9, 10, 23, 24, 26–29, 67, 68,
    79, 93–97, 103, 107–124,
    126–133, 135, 137, 149,
    150, 155, 156, 162, 177,
    178, 185, 211–218, 220,
    222, 223, 296
  Critique
    Third Man Argument,
      120, 178, 213
  Dialogues
    Laws, 118–120
    Meno, 24, 27, 94, 96, 107,
      109, 110, 116, 296
    Phaedo, 107, 110–113,
      116, 211, 212, 296
    Philebus, 118, 126, 127,
      131
    Republic, 113, 114
    Sophist, 123, 124
    Symposium, 107, 112, 113,
      116, 211, 212
    Theaetetus, 97, 118–123,
      131
    Timaeus, 117, 118, 127,
      128, 131, 214, 215, 296
  Epistemology
    Likely Story (Eikos
      Mythos), 117, 128, 215,
      296
  Metaphysics
    Receptacle (Khōra), 127,
      128, 215

Popper, Karl, 17, 101, 114, 229,
    230, 241, 273
Popperian Protocol, *see*
    Theorems, T1:
    Popperian Protocol
Pragmatism, 81
Pre-Platonic, 3, 23–25, 28, 29, 31,
    37, 42, 48, 51, 56, 79, 85,
    94, 104, 107, 108, 112,
    116, 118, 133, 136, 145,
    147, 173, 187, 188, 195,
    211, 215, 217, 221–223,
    230, 231, 235, 247, 249,
    256, 257, 265, 275, 293,
    306
Process Philosophy, 15, 33, 136,
    208, 232
Prohairesis
  aesthetics, 168
  ethics, 99
  navigation, 94, 99
Psyche
  Aristotelian Hylomorphism,
    137
  Platonic Dualism, 109, 110
  Socratic Therapy, 23
  Stratified Architecture
    (NPN), 9, 15, 21, 79, 89,
    90

R

Railton, Peter, 54
Region R, *see* Logic of
    Distinction, Interstitial
    Ground

S

Second-Best State, 130, 149, 155, 156, 255
Socrates (Historical Figure), 9, 21, 23–29, 93, 94, 110, 117, 120, 121, 132, 235, 265, 295–297, 299
Socratic Method, 86, 94, 113, 167
Somatic Dhamma, 257
Somatic logos, 5, 26, 41, 55, 60, 67, 71, 76, 80–82, 91, 101, 111, 124, 156, 171, 172, 191, 198, 225, 227, 241, 258, 259, 262, 304
    epistemology, 79
    First Principle, *see* First Principles, FP7: Somatic Logos
    navigation, 98
Somatic Present Corollary, *see* Corollaries, C2: Somatic Present
Spencer-Brown, George, 35, 180–182, 185, 277
Spinoza, Benedict de, 7, 35, 37, 49, 180, 182, 203, 280
    Determinatio est negatio, 37, 180, 280

T

Taiji (Daoism), 179, 248, 249, 251, 256
Thales, 10–12, 20, 293, 294
The Cut, *see* Logic of Distinction, Adikia (The Cut)
Theorems
    T1: Popperian Protocol, 231
    T2: Status of Formal Truths, 88, 196, 235
    T3: The Necessary Distortion, 84
    T4: Ethical Isomorphism, 59, 60, 244, 291
    T5: The Entropic Mandate, 15, 58, 59, 64, 158, 198, 199, 206
    T6: Life-Agency Isomorphism, 61
    T7: The Entropic Asymmetry, 54, 289
Thermodynamics
    Life-Agency Isomorphism, 200
Tyrant (Political Pathology), 115, 116

U

Uncertainty Principle, 232
Unmoved Mover, 141, 142, 220, 222, 223

V

Value
- Functional Alignment, 23, 54, 60, 65, 90, 122, 123, 146, 162, 197, 198, 247, 288, 301

Varela, Francisco J., 52

Vlastos, Gregory, 26, 29, 95, 120, 204

W

Whitehead, Alfred North, 31, 104, 136, 138, 139, 219, 232

Wu Wei (Effortless Action), 254, 255

Wuji (Daoism), 179, 248–250, 256

Y

Yin and Yang, 15, 38, 100, 247, 249

Z

Zero Principle, *see* First Principles, ZP: Zero Principle

About the Author

Eli Adam Deutscher is a philosopher, systems architect, and the founder of Neo-Pre-Platonic Press. His work begins from a functional crisis: the inability of modern frameworks to provide coherent, actionable answers to fundamental questions of knowledge, value, and purpose.

Trained in the rigorous, applied disciplines of engineering and economics at the **Colorado School of Mines** and tested in the complex, strategic arena of consultancy, Deutscher approached philosophy not as a historical debate but as a **first-principles design problem.** *Neo-Pre-Platonic Naturalism* is the result: a complete, deductive meta-structure that re-grounds human meaning in the constitutive logic of a lawful cosmos.

He writes for those who require a philosophy that is not merely believed, but **built—and used.**

A Note on the Production

This volume was published by **Neo-Pre-Platonic Press** (an Imprint of the NPN System) in Winter 2025.

Book Design and Typesetting executed by Eli Deutscher using LaTeX and Pandoc. © 2025 Eli Deutscher

Body text is set in Libertinus Serif with 11pt.

The cover was designed by the Author. © 2025 Eli Deutscher

The NPN Logo is a Trademark™ of Neo-Pre-Platonic Press.

www.ingramcontent.com/pod-product-compliance
Lightning Source LLC
LaVergne TN
LVHW092013090526
838202LV00026B/2632/J